EUROPEAN TECHNOLOGY
FOR OBTAINING ENERGY FROM SOLID WASTE

EUROPEAN TECHNOLOGY FOR OBTAINING ENERGY FROM SOLID WASTE

Edited by D.J. De Renzo

NOYES DATA CORPORATION
Park Ridge, New Jersey, U.S.A.
1978

Library of Congress Catalog Card Number: 78-61896
ISBN: 0-8155-0730-5
Printed in the United States

Published in the United States of America by
Noyes Data Corporation
Noyes Building, Park Ridge, New Jersey 07656

Foreword

The recovery of energy through the combustion of municipal solid waste is by now a well recognized and established technique for conserving energy. Combustion units can produce electricity, hot water for domestic use and steam for district heating and for industrial processes, or for the drying of sewage sludge.

Western Europe is definitely the leader in this field. Energy recovery from waste began, in a small way, in France, Denmark, and Italy before the Second World War. During the 1950s it was introduced on a large scale in West Germany, France and Holland. The largest single furnace is located in Paris-Ivry and has a capacity of 50 metric tons per hour, while the largest European plant near Rotterdam has 6 units of 20 metric tons per hour.

Much can be learned from these European waste-to-energy systems. For each country a national overview is given, followed by a description of particularly significant developments and case histories: household sorting and collection methods, combustors, furnaces, incinerators, air pollution control, latest plant designs, operation and economics.

A summary of key findings is also included. Serious study of this book should make it possible to reveal and identify solid waste/energy processes (both existing and under development) that appear to offer potential advantages over processes currently employed in the U.S.

Because the information in this book is taken from multiple sources, it is possible that certain portions of this book may disagree or conflict with other parts of the book. This is especially true of monetary values and opinions of future potential. We chose to include these different points of view, however, in order to make the book more valuable to the reader.

Cost figures provided are those given in the report cited, the date of which is always given. When the dates of the cost figures themselves are given, we have included them.

Advanced composition and production methods developed by Noyes Data are employed to bring these durably bound books to you in a minimum of time. Special techniques are used to close the gap between "manuscript" and "completed book." Industrial technology is progressing so rapidly that time-honored, conventional typesetting, binding and shipping methods are no longer suitable. We have by-passed the delays in the conventional book publishing cycle and provide the user with an effective and convenient means of reviewing up-to-date information in depth.

Maps and a tabulation of the European systems are included. Some of the tables and maps have been reproduced by photo offset directly from the original typeset reports, a technique which saves time and prevents errors from occurring.

The Contents and Subject Index provides easy access to the information contained in this book which is based on several studies produced under contract for EPA and ERDA (now DOE). These primary sources are listed at the end of the volume under the heading, Sources Utilized. For the convenience of the user two conversion tables are appended: one is a monetary conversion table valid for the time span covered in the reports and the other compares metric engineering units with the English units still prevalent in the U.S.A.

Contents and Subject Index

European Developments in Waste-to-Energy Systems

Material for this chapter has been based upon a report prepared by W. David Conn of the University of California, Los Angeles, California (PB-270 219).

INTRODUCTION

Current (summer, 1975) priorities in the United States with respect to energy and materials recovery from municipal solid waste include (a) the use of refuse as a supplementary fuel, (b) pyrolysis, and (c) resource recovery. Reported here is a study which set out to determine whether the same priorities apply in Western Europe and to identify solid waste/energy processes there (both existing and under development) that appear to offer potential advantages over processes currently employed in the United States.

The study included a review of published and unpublished literature, telephone and letter contacts, and selected visits by the principal investigator.

The information is presented as received; time and resources have not permitted an independent verification of its accuracy and completeness. Most information was collected prior to September 10th, 1975.

SUMMARY OF KEY FINDINGS

Landfill generally remains the preferred method of disposal in Western Europe where suitable sites are available, but these are becoming increasingly difficult to find, particularly near the large cities.

In some countries, composting is considered viable, but only on a limited scale. There is a demand for the product from vineyards, etc.

Incinerators are in common use, especially in urban areas, but costs have increased so rapidly that several governments are discouraging the construction of new plants. Increasingly stringent air pollution regulations necessitating the use of gas scrubbers (e.g., in West Germany) can cause the costs to rise even higher.

Provision for heat recovery is made in many incinerators, especially the larger plants. The technical difficulties (notably corrosion) are now reasonably well understood, although unexpected problems can still arise. The problem of matching supply with demand remains a significant barrier to heat recovery, especially for electricity generation (which is generally considered worthwhile only in the very largest plants). The recovery of heat for district heating and/or industrial use is preferred, particularly when an incinerator can provide the base load throughout the year. Rising fuel costs are making the economics of heat recovery more attractive.

Most countries are interested in the possibilities offered by the new technologies of materials/fuel separation, pyrolysis, etc. On the whole, they are awaiting "hard data" from the United States but in some places development is proceeding independently.

Countries in which front-end materials/fuel separation processes are being developed include England, Italy, Spain, Sweden, and West Germany. Of particular interest is the burning of refuse-derived fuel in cement kilns.

Countries in which pyrolysis processes are being developed include Denmark, England, Netherlands, Sweden, and West Germany.

Some work has been done on materials recovery from postincinerator residues (notably in England and France) but there seems to be little enthusiasm from incinerator operators for installing a full-scale system.

Several countries are experimenting with household sorting and separate collection systems, the discouragement of nonreturnable containers, etc.

BELGIUM

National Overview

Other than separate collection, there are no systems for the recovery of energy/materials from municipal solid waste operating in Belgium at the present time although a cryogenic process for separating 99% pure ferrous metal using liquid nitrogen is reputedly being tested in Liege by George and Sons (no further information was obtained). However, careful study is being made of recovery possibilities.

DENMARK

National Overview

The principal government agency concerned with solid waste management is the Miljostyrelsen (Ministry of Environmental Control).

Approximately 60% of refuse in Denmark is incinerated in fairly modern incinerators, one-third of which have been constructed in the past 5 to 6 years. Two large plants serve the Copenhagen area (representing 2 million out of the total Danish population of 5 million). Emission standards can readily be met using electrostatic precipitators so that pollution control is not a major constraint on incineration (as it is in Germany). No financial assistance is provided by the central government for the construction of disposal facilities.

Many of the incinerators recover heat; for example, one supplies heat to a power station and another will soon do so to a new 1,000 bed hospital (which is not yet completed). There is a tradition of district heating based on hot water rather than steam, because the Danes prefer to operate their incinerators at lower temperatures and pressures than those used in steam systems, in order to reduce maintenance problems. However, district heating is no longer considered as economically attractive as before, and there has been a tendency to build newer incinerators, especially the smaller ones, without heat recovery.

Although the Minister has made a statement to Parliament in favor of resource recovery, etc., no government support has been forthcoming. In some municipalities, refuse is separated by households into three bags (metal and glass, paper, and the rest); this is organized by the local authorities in collaboration with consultants and with the company selling the bags. (This is an offshoot of a Swedish program; see below.) For example, in Birkerod (population 20,000) success is claimed as there is 90% participation. However, it may be that waste is increased, as families are happy to have three bags instead of one. Furthermore there are problems in marketing the recovered materials.

There is a Danish tradition of using returnable beverage bottles, and this still holds. Nonreturnable bottles are relatively rare, although there are quite a few cans. The government has made an agreement with the Danish breweries to restrict nonreturnable containers to 2 to 5% of the total market. A large supermarket has experimented with returnable wine bottles, but was not successful because of the diversity of bottles involved.

Notable Development

Development by Pollution Control Ltd: A pilot/demonstration pyrolysis plant utilizing the "Destrugas" process has been constructed at Kalundborg (about 100 km west of Copenhagen) by Pollution Control Ltd. The company is a limited partnership formed by Messrs. Karl Kroyer (inventor-and engineering-company) and Superfos Ltd. (the largest industrial chemical company in Denmark). Figure 1.1 is a schematic of the plant.

The pilot plant has been proven at slightly over 5 metric tons/24 hours. A single retort is used which would not be scaled up in a full-scale plant (as the heat transfer characteristics must be maintained); instead, banks of retorts would be used. The company's target is to construct plants of around 100 to 500 tons/day (rather smaller than some of the United States facilities).

After pulverizing in a hammermill to minus 100 mm (larger particles may be acceptable, especially if they are flexible), the waste materials are fed into a vertical, indirectly heated retort tube, where they are pyrolytically decomposed in the absence of oxygen at temperatures up to 1000°C (Figure 1.1). The process produces a sterile char, free of biologically decomposable material, and a gas with a composition resembling that of coal gas. Part of the gas is fed back into the system to be used as fuel; the remainder is available for use elsewhere. It can be mixed with natural gas (giving a mixture with a calorific value substantially less than that of natural gas alone) or, more likely, it can be used industrially. The evaporated water from the waste materials condenses in the scrubber system and is led to the quench tank to cool the char. Provided that the moisture content of the incoming wastes is no more than 40%, all of the water is absorbed by the char (otherwise there might be some surplus, requiring treatment). Atmospheric emissions from the stack are claimed to be less than 10% of the emissions from a modern incinerator of comparable size.

Instead of using the char to absorb the water (which would therefore require alternative treatment) it could be used as a fuel, as a filter (it could possibly be converted into a high grade activated charcoal), or in combination with sewage sludge as a soil conditioner.

Pollution Control Ltd. has established links in various countries, including West Germany, England, Japan, Norway and New Zealand. In Japan, a demonstration plant similar to the Kalundborg plant is to be built under license in the city of Hitachi (design work currently being started). Prospects in Japan are considered very good, and negotiations are proceeding with several local authorities.

In West Germany, a joint venture is being established with Wibau Matthias and Company. The Bavarian Ministry requested a 3 to 5 week demonstration at Kalundborg with independent monitoring; the West German government then took over the research and contributed financial support. The trial was run by the University of Stuttgart over a six week period in summer, 1974, using four different kinds of waste (Danish refuse, German refuse, refuse plus sludge, and refuse plus waste tires). The results are still confidential but are claimed to support those of Pollution Control.

The Bavarian Ministry requested proposals for a 180 tons/24 hour facility to be built in the Munich area. In competition with Torrax, Union Carbide, etc.,

FIGURE 1.1: PC-PYROLYSIS PLANT, KALUNDBORG, DENMARK

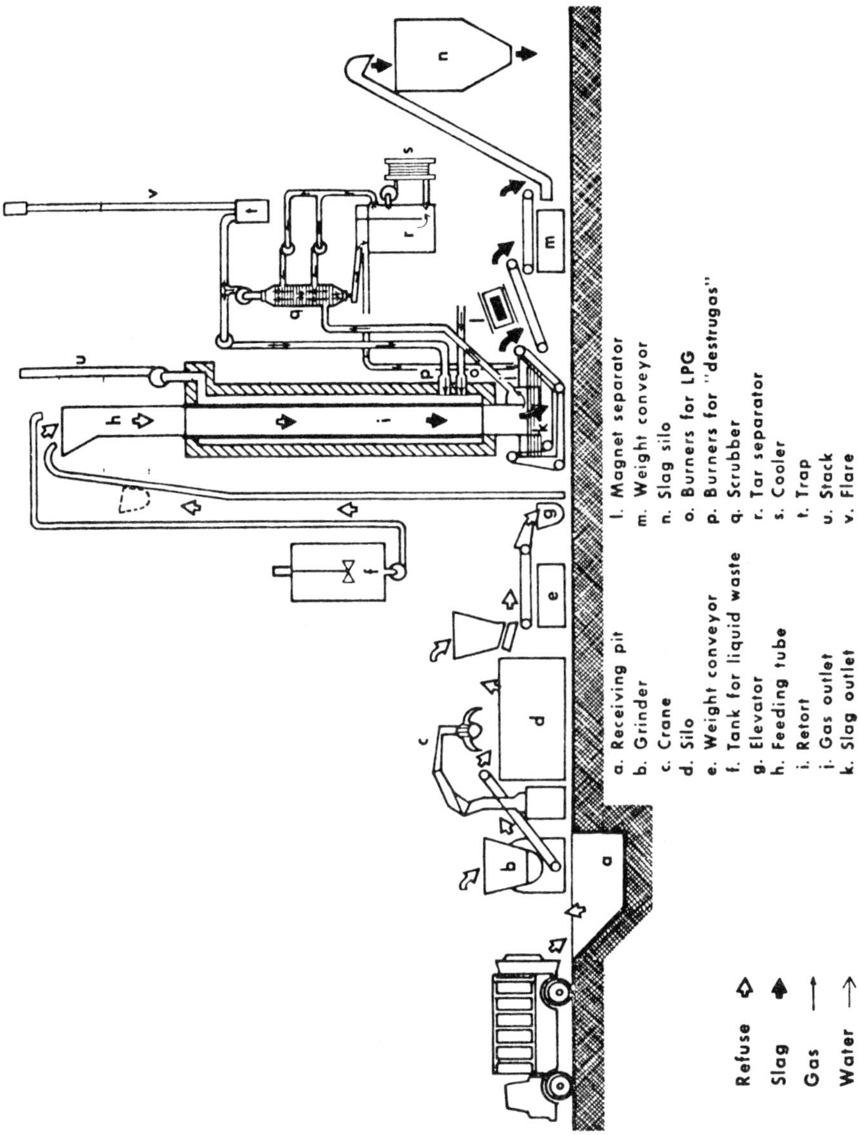

Pollution Control won the right to negotiate, and is currently submitting a quotation for detailed engineering studies. The intention is to initially install furnaces handling 90 tons/24 hour day, but to increase this capacity after one year if the operation is successful. The plant should run for at least 8,000 hours/year, leaving about 30 days for outages, but it is claimed that the plant could run longer. The construction is such that the whole heated zone should be maintained continuously at full heat. Based on experience with coal gasification plants, a retort should have a 10 to 15 year lifetime; repairs and maintenance should be mainly confined to the handling facilities, fans, etc.

The Bavarian Ministry has not received financial support from the West German Government which may instead support the construction of a pilot plant to be owned by the Berlin Solid Waste Authority in collaboration with the Technical University of Berlin. If constructed, the plant would have a capacity of at least 15 tons/day (possibly 30 tons/day). A contract for the predesign stage is currently being negotiated.

Estimates of costs have been made carefully and are thought to be accurate to within about 10%. The capital costs include design and construction fees, etc., based on a normal site but do not include the cost of the site itself nor the provision of services, fences, etc. For a 120 tons/24 hour plant, total capital and operating costs would amount to 106.8 Danish Kroner (kr) per ton (made up of capital costs, 66.8 kr; repair and maintenance, 9.5 kr; operation, 10 kr; salaries and wages, 20.5 kr). Assuming an income of 45 kr per ton from the sale of surplus gas, ferrous metals, and carbon char, the net cost per ton would be 61.8 kr. The cost decreases as the plant size increases, up to a 360 tons/24 hour plant for which the gross cost per ton would be 87.8 kr (net, 36.8 kr).

ENGLAND

National Overview

The principal government agencies concerned with energy/materials recovery from municipal solid waste are (1) the Department of the Environment and (2) the Waste Disposal Authorities (the County Councils).

The three United States priorities are thought generally applicable, but two important factors must be borne in mind: (1) local government recently underwent a major reorganization and many of the newly constituted Waste Disposal Authorities are still in early stages of planning; and (2) the Control of Pollution Act 1974, which has major provisions affecting the disposal of municipal solid waste (e.g., in requiring the Waste Disposal Authorities to prepare plans and issue disposal licenses), has not yet been implemented owing to the unavailability of financial assistance from the central government. The situation is therefore somewhat transitory, although some interesting developments are occurring.

The Department of the Environment (DOE) until 1970 was required to approve solid waste projects where loan sanction was sought by local authorities but this no longer applies. It now sets a figure each year for capital investment, leaving the local authorities discretion over how it is spent. The DOE gives "advice on how to select options" and this is generally to use sanitary landfill whenever possible (as it is considered the least expensive method of disposal). Where landfill sites have been unavailable in the past, many local authorities have built direct incineration plants (some 20 to 30 authorities built them prior to 1970) but costs have doubled in the past four years or so and the DOE now advises against their construction. Present advice is to hold off heavy investment in conventional equipment until the new processes (i.e., pyrolysis, fuel/materials separation, etc.) have been fully examined.

A Waste Management Advisory Council has been established and is currently looking into resource recovery (although no actions have yet been taken). In order to stabilize waste paper markets, a buffer stock scheme is being considered, in which industry would be paid to hold stocks. Packaging is being examined, and three industry reports (on glass, metals, and plastics) have been published.

Notable Developments

Research at the Warren Spring Laboratory (operated by the Department of Industry): An early project on the extraction of nonferrous metals from incinerator clinker (Figure 1.2) has been completed. Research (funded by the DOE) is progressing on (1) fuel/materials separation, (2) pyrolysis, and (3) the pneumatic conveyance of pulverized refuse.

(1) In the fuel/materials separation project, an attempt is being made to separate as many components as possible, notwithstanding present market conditions, which are thought likely to change at any time. However, any given commercial plant would be designed according to the particular circumstances then existing. A pilot system has been constructed at Warren Spring (Figure 1.3). A key feature of the process is the absence of a hammermill or other shredder at the front end (allowing appreciable energy savings). Other features include a ballistic drum (rather than plate) separator (to prevent nylons, etc., from becoming caught) and a thrower/separator which relies on a combination of aerodynamic, impact, and sliding friction effects to separate a low density product (mainly small pieces of paper, vegetable and garden wastes, and low density plastics, etc.) and a high density product (mainly dense vegetable matter, broken glass, bones, etc.).

No attempt is made to color-sort the glass, because unless the white fraction can be made clean enough for the manufacture of flint glass (which is currently thought unlikely), it is not thought worthwhile to separate at all. Some experimentation with glass color sorting has been conducted. The color-mixed froth floated product currently produced contains more carbon particles than does the United States product, giving it a gray cast. The carbon burns off

FIGURE 1.2: WARREN SPRINGS PROCESS FOR IMPROVED METAL RECOVERY FROM INCINERATOR CLINKER

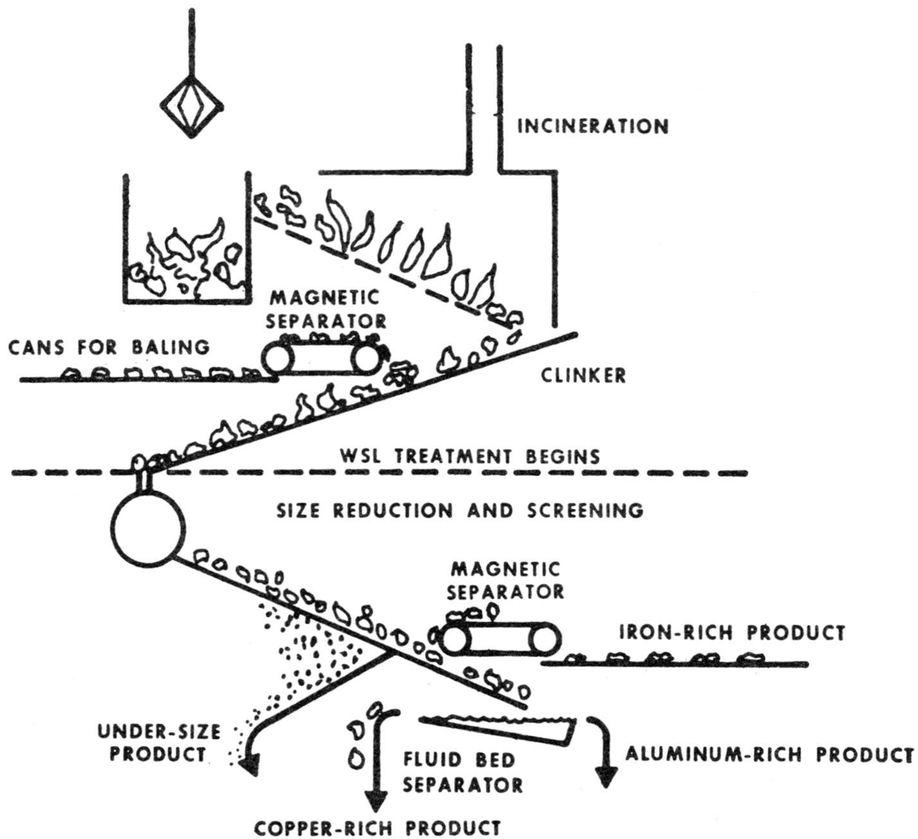

FIGURE 1.3: WARREN SPRINGS LABORATORY-SCRAP AND WASTE SECTION PHYSICAL SORTING OF DOMESTIC REFUSE

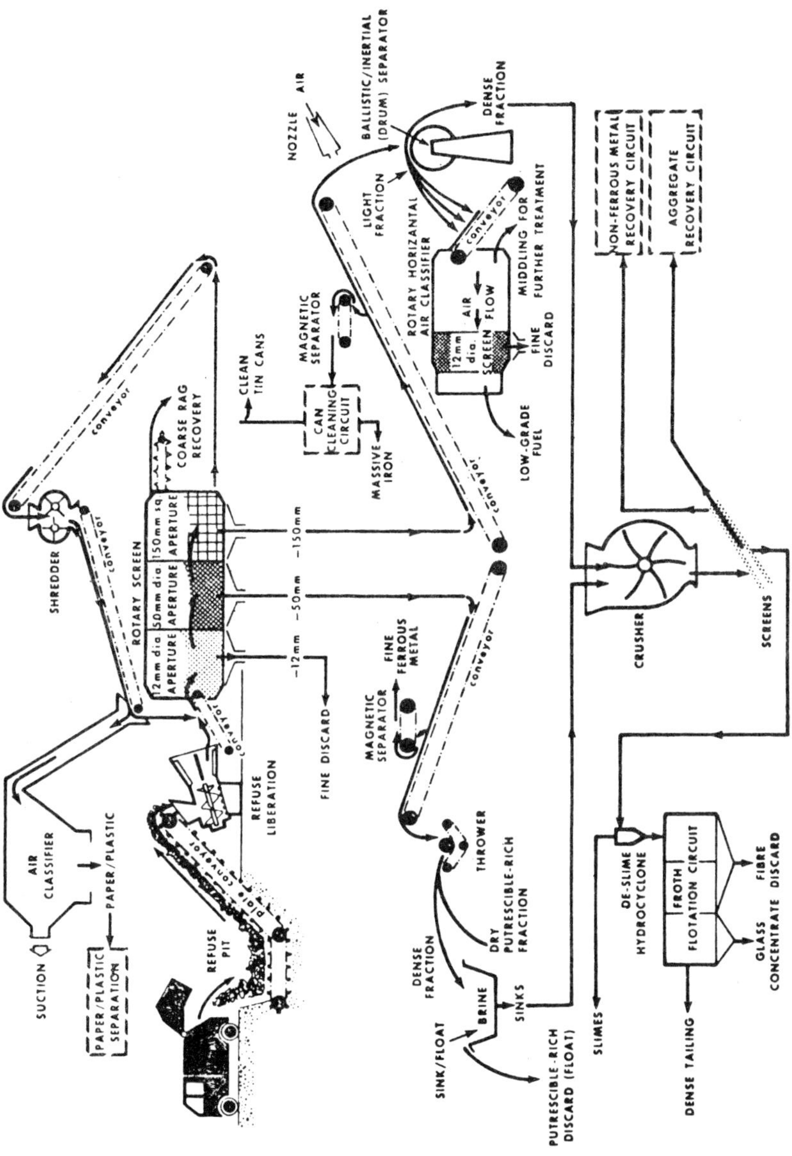

during glass melt, thereby providing some fuel value; as a result, British manufacturers have suggested that the glass should carry a premium price compared with material containing less carbon. Some 50% by weight (or about 20% by volume) of the incoming refuse forms a residue that remains for disposal. The froth flotation circuit produces a contaminated liquid effluent.

An economic assessment of the process is currently being prepared. It is tentatively considered that a 200 tons/day plant could break even (including the cost of residue disposal). The process is about ready to go commercial, and discussions are being held with local authorities and some industrial companies. The first plant will probably produce mainly a fuel, together with cans, nonferrous metals, and glass. [The construction of two separation plants (each 300 tons/day) has since been announced, one to be built for South Yorkshire, the second for Tyne and Wear.]

(2) The project on pyrolysis has proceeded in three stages (Figure 1.4). Following bench-scale tests, a retort with indirect heating was built and operated at temperatures of 600° to 1000°C to provide background data as well as products for further examination. The major disadvantage of this design stems from the poor thermal conductivity of the refuse, necessitating a considerable retention time for peripherally supplied heat to conduct to the center of the charge. The second stage of research utilized an induction heating system; the incoming refuse was mixed with 4 cm diameter steel balls and passed down the reactor tube and through an induction coil. The advantage is that in some instances, working temperatures can be reached in one-tenth of the time required in an externally-heated retort, but the circuit is complicated; in a full-scale plant, capital and operating costs would be high. In order to be self-sustaining, the energy from the char and the gas would be needed to generate electricity. The process might be economical on a large scale (the size of the coil being a limitation) or in special applications (e.g., in an enclosed system for the disposal of radioactive waste).

The third stage process utilizes a cross-flow heating system in which a combination of fuel from the char and some gas is used in a heat-exchanger to heat and pyrolyze the incoming refuse. Based on a calorific value for the incoming refuse of 4,500 Btu/lb, the net energy output (as excess fuel gas) is estimated at 40% of the input energy. The costs are currently being worked out; the heat exchanger and fan are particularly expensive items, but the overall costs for a 100 tons/day plant are expected to be about two-thirds those of an equivalent-sized incinerator. The pyrolysis plant would be roughly equivalent in size to the gas-conditioning plant of the incinerator. Problems might arise in the treatment of the liquid effluent (depending on the conditions, about 2 to 20 gallons/ton of very highly contaminated effluent would be produced, which would be treated using activated sludge) and in the preliminary gas cleaning prior to recirculation (as some organic material and carbon might be deposited due to cracking of the gases; results so far suggest not much of a problem, and the remedy would be scrubbing with an inexpensive scrubbing medium). Air pollution should not be a problem; the combustion gases from

FIGURE 1.4: WARREN SPRING LABORATORY: PYROLYSIS

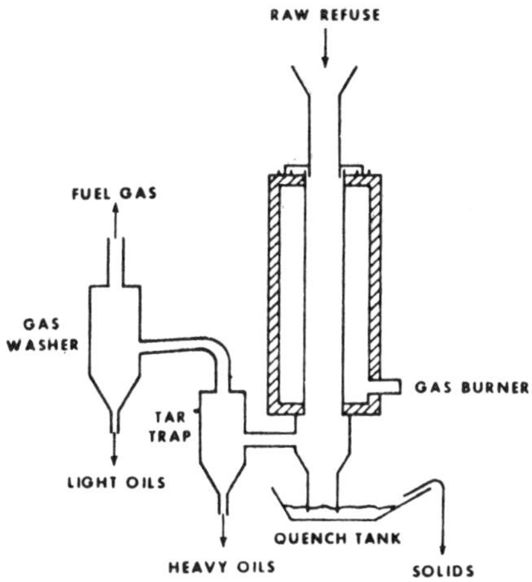

STAGE ONE: INDIRECTLY FIRED RETORT

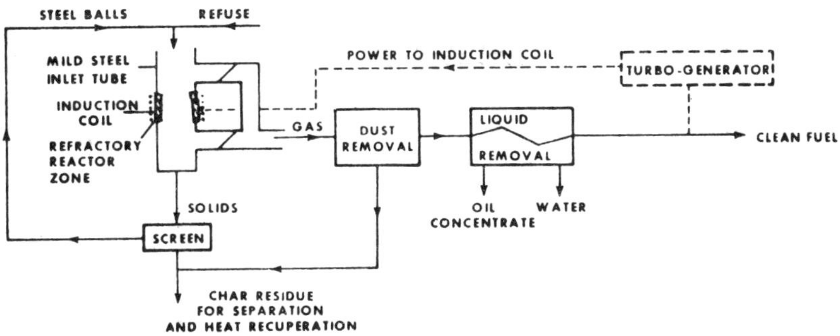

STAGE TWO: PYROLYSIS BY INDUCTION HEATING

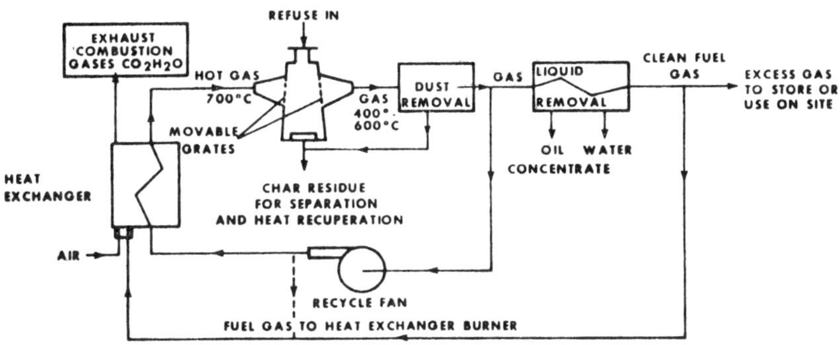

STAGE THREE: PYROLYSIS BY HOT PRODUCT GAS RECYCLING–CROSS FLOW PRINCIPLE

the heat exchanger should be clean unless the char is also used as a fuel, in which case there might be some particulates. However, there should be less problem than with pulverized fuel combustion, and an electrostatic precipitator would probably be unnecessary.

(3) The project on the pneumatic conveyance of pulverized refuse utilizes an 8″ diameter line and has so far been proven over a distance of 350′ (intending to increase to 1,000′). The refuse is injected into the line by means of a screw mechanism, and is conveyed in "plugs" with air cushions in between. The pressure is low (2½ psi), there is no filtering requirement, and the system is inexpensive to operate. The system may be tried at the IMI facility (see below).

Developments at the Greater London Council (GLC): The GLC handles just under 3 million tons of refuse per year, or about 11,000 tons per working day. Its policy is to employ several methods for disposal rather than relying on only one, although it looks for facilities with capacities of no less than about 600 tons/day.

Sanitary landfill remains the preferred method when available, but the shortage of sites is critical. Membranes are being considered at marginal landfill sites, but the GLC is also looking at pyrolysis, fiber reclamation etc. Although the GLC operates a major new heat recovery incinerator at Edmonton (see below), it is unlikely to build another incinerator in the foreseeable future because of high costs and the difficulty of finding appropriate sites.

In partnership with Warren Spring Laboratory and Foster Wheeler (equipment manufacturer) the GLC may test out the Warren Spring pyrolysis system still being negotiated). It may also send ½ million tons/year of refuse to the APCM system (see below).

The Edmonton incinerator was designed for an average input of 1,800 tons per day, five days per week, with a plant throughput of 1,300 tons every 24 hours and a peak of 1,670 tons. The electric power output was designed to be within the range of 25 MW and 35 MW but with a peak of 45 MW according to calorific value and throughput of refuse type roller grates; the boilers are single drum water tube boilers with partially water-cooled combustion chambers.

Commissioning of the plant took place during 1970/71. Teething troubles were aggravated by unexpected failures due to boiler tube erosion and corrosion. The failures were extensively investigated, and various modifications were made at the end of 1972. The plant did not become fully operational until April 1974. In the subsequent 12 months, 378,000 tons of refuse were incinerated (averaging more than 1,000 tons/day) and 162 units of electricity were generated. Approximately 94% of the planned boiler availability (i.e., 4 out of 5 boilers in continuous use) was achieved, but only 77% of the refuse throughput. The difference was due to limitations placed on incinerating capacity by higher-than-acceptable steam temperatures at the superheaters when the furnaces worked at full load (new modifications will cope with this problem).

The total capital cost of the plant, including all major modifications to the boilers, was approximately £13.5 million. After deducting an income (from electricity, baled ferrous metal, and furnace ash) of £788,000, the net operating expenditure for the 12 months from April 1974 was £2,160,540, including debt charges; this represented a net unit cost of £5.72 per ton. These costs are considered "encouraging" as the costs of alternative systems of long distance haulage to landfill sites are thought likely to rise at a greater rate than those of the Edmonton plant.

Development by Associated Portland Cement Manufacturers (APCM): APCM has been working on a system for recovering fuel from refuse for about 5 years. A contract has been signed with Wiltshire County Council for the delivery of 60,000 to 80,000 tons/year of crude refuse. This will be pulverized in a hammermill (probably), passed through a magnetic separator to remove the ferrous, screened in a trommel, and then pulverized again in a shredder or hammermill (trials still under way); experiments have shown that air classification is not necessary. The resulting material (about 30 to 50 mm diameter) will be conveyed 1,000′ and fired directly into a rotary cement kiln. The residue will be absorbed into the resulting cement, and careful tests have shown that there are no adverse effects on the properties of the cement. The use of 4 tons of crude refuse will save 1 ton of clay or alternative raw material (presumably as a filler).

An abrasion problem is anticipated and is being designed for. The kiln temperature is 1450°C. The resulting gases will be scrubbed in a heat exchanger; hydrogen chloride will be neutralized, leaving a minute concentration of chlorine in the cement (difficult even to detect). Chlorine will also be present in the particulates trapped by the electrostatic precipitator, but this should pose no problems. Even if the plastic content of the refuse rises to 5%, with one-fifth of this being PVC, no problem is anticipated. All drainage from the building (inside and out) will go into a well and be used to blend slurries which then go into the kiln (therefore, no problems).

The gross calorific value of the refuse is estimated at about 4,000 Btu/lb, depending on conditions. Some heat will be lost up the chimney, some will be used to dry the refuse (about 30% moisture content), and some to heat up incoming air. APCM seeks to replace about 10% of the fuel equivalent with refuse.

Start-up is expected by August/September, 1976. Wiltshire will pay a dump fee; the 10 year contract allows for variations if fuel costs increase, etc. APCM is currently in active negotiation with 4 other country councils (including the GLC). If all present cement works were converted, it is estimated that between one-sixth and one-seventh of the total UK municipal waste could be absorbed. It might also be possible to inject more unpleasant wastes into the pneumatic line (especially organics, which would easily be destroyed in the kiln). The process is provisionally patented in the UK and a patent is pending in the United States.

Developments at the West Midlands County Council (WMCC): The WMCC disposes of approximately 1 million tons of solid wastes per year, mostly from domestic and commercial premises. Approximately 50% is disposed of by landfill, and 50% by incineration. A new incinerator has recently been opened at Coventry, and another is being built in Birmingham.

The Coventry Incinerator plant, costing about £4,138,000 at November 1974 prices (mechanical and electrical equipment, £2,300,000; civil engineering and building works, £1,500,000; land acquisition, design and supervision of construction, £338,000) has three boilers having 12 tons/hour capacity. Two boilers normally operate on a continuous basis. The boilers have water wall combustion chambers as well as shell and tube heat exchangers for heat recovery. The resulting steam is currently used in the plant to drive steam turbines, the excess steam being condensed in air cooled condensers. Within the next 12 months a heat transfer station will be completed and the heat will be used to provide hot water to an adjacent factory.

WMCC will soon be supplying up to 15,000 tons of refuse per year to the IMI waste utilization project (see below).

Development at Imperial Metal Industries Ltd. (IMI): IMI operates its own power plant at Birmingham with a steam raising capability to satisfy the factory's entire electrical power requirements as well as providing low pressure steam for space heating and process needs. The boilers currently burn coal, heavy fuel oil or natural gas in any combination. A refuse preparation plant is being constructed at which refuse will be shredded in a Tollemache vertical shaft pulverizer (100% minus 150 mm and 80% minus 50 mm). Ferrous scrap will be separated magnetically. The remaining refuse will be conveyed by container one mile to the power plant, where it will be fired in one of three modified boilers (Babcock and Wilcox water tube boilers fitted with chain grate on which there is an established coal fire). A 50/50 refuse/coal mixture is thought achievable, with the possibility of increasing the refuse fraction to 75%. The plant includes grit arrestors and an electrostatic precipitator, and air pollution problems are not anticipated.

IMI originally intended to include air classification of the refuse but further study convinced them of difficulties that were apparently borne out by their visit to St. Louis. The difficulties are the loss of potential burning power in the heavy fraction, and mechanical problems (e.g., blockages).

In the future, IMI might employ the pneumatic conveyance system developed at Warren Springs to transport the shredded refuse from the processing station to the power station.

IMI has a five year initial contract with the West Midlands County Council which will pay £2 per ton for the first 15,000 tons disposed of in each of the first two years of the contract. Any additional refuse in the first two years will be disposed of without charge. It is uncertain whether charges will be made after the first two years. IMI is interested solely in the fuel value of the refuse (and not in its disposal) and the operation should be self-supporting.

Development at the Open University: Dr. Andrew Porteus is continuing his work on the acid hydrolysis of refuse (reported, for example, in Drobny, N.L., H.E. Hull and R.F. Testin, *Recovery and Utilization of Municipal Solid Waste*, Report SW-10c, Environmental Protection Agency, 1971). Alternative products include ethanol or edible proteins.

Despite analyses by Porteus indicating that the process is economically viable, little interest has been shown by local authorities. This may be partly because of the high proportion (50 to 60%) of the wastes left for disposal (i.e., the process is not a disposal process per se, but rather a recovery process).

Other work on hydrolysis is reportedly being done at the University of Wales by Professor D.E. Hughes and Dr. C.E. Forster.

FRANCE

National Overview

The principal government agencies concerned with energy/materials recovery from municipal solid waste are (1) Ministère de la Qualité de la Vie (Ministry of Quality of Life) and (2) Ministère de l'Industrie et de la Recherche (Ministry of Industry and Research), as well as the Delegations for the Saving of Raw Materials, for the Saving of Energy, and for New Forms of Energy.

A new law (Law No. 75-633, 15th July 1975) is similar in philosophy to the U.S. Resource Recovery Act. Furthermore, it provides the possibility of controlling methods of manufacture and of constraining certain activities (e.g., prohibiting or taxing nonreturnable containers); where recycled materials are not significantly different from virgin materials, it prohibits advertising on the basis of virgin content. It is not clear when and to what extent these provisions will be implemented.

The central government gives the local authorities up to 20% support for disposal facilities, but until recently this could be applied only to the basic plant (e.g., an incinerator) and not to associated recovery equipment (this is now changing). Old fashioned ideas, fluctuating markets, and product specifications requiring virgin materials are among the factors that have militated against recovery.

At the present time, somewhat more than 30% of the municipal solid waste produced in France is disposed of by incineration, about 10% by composting, and about 60% by landfill (including 15% by sanitary landfill). Composting is still considered viable under certain circumstances, as orchards and vineyards are good markets (the plastics content can pose an aesthetic problem, although it may be good for binding the soil). A mechanized plant with capacity of 60,000 tons/year (175 tons/day) was opened in 1974 in Reims.

Approximately 60% of the incinerator plants have heat recovery, mostly in the form of district heating (although some electricity is generated, e.g., in Paris). It is considered worthwhile to install recovery in plants of capacity exceeding about 200 tons/day. Provision is usually made to meet only a proportion (i.e., about 5 to 10%) of the demand for heat during the winter, so that the supply and the total demand are well matched in the summer. However, incineration is now so expensive that the central government is discouraging its selection. Despite low interest charges (about 7 to 8%) and the 20% capital cost contribution, it can cost a local authority up to 60 to 70 francs/ton (approximately 50/50 capital/operating costs). Compare composting at about 35 francs/ton (mechanized) or about 25 francs/ton (nonmechanized) (assuming and including credit for sales), or landfill at about 15 francs/ton (unshredded) or about 20 francs/ton (with shredding).

Pyrolysis, fuel separation, etc., are being examined. A small pilot plant (based on the Garret pyrolysis process) was constructed by Sodeteg Engineering about two years ago near Rouen, but there were problems in maintaining continuous operation and the project was abondoned. Work is also being done at Orleans (see below).

In materials recovery, a committee is considering means to stabilize prices, possibly by maintaining a flexible stock. Work on deinking is also proceeding. Glass manufacturers have expressed themselves ready to accept all recovered glass at a firm price equivalent to the price of alternative raw materials (possibly in order to avoid legislation on returnable bottles). The central government may give grants to several towns to promote separate collection of paper, glass, and plastic; a new process called ENERGY has been developed to reuse mixed waste plastics (even if soiled) in making a product which can be reformed.

Notable Developments

Heat Recovery Incinerators in the Paris Area: There are incinerators at Issy (555,000 tons/year), Ivry (630,000 tons/year), and St. Ouen (360,000 tons per year). The plants serve a population of about 5 million, and handle about 1,650,000 to 1,700,000 tons of refuse per year, together with about 50,000 tons/year of industrial waste. While St. Ouen produces just steam for district heating, Issy and Ivry produce both electricity and steam for district heating in flexible proportions, depending on demand, etc. The electricity is sold to Electricité de France (at a price maintained artificially low) and the excess steam is sold to Chauffage Urbain, which has a 219 km district heating network in Paris (also at an artificially low price, namely one-half of the cost of producing steam from an alternative fuel). Experience has shown that the higher the steam pressure, the greater the problems (thus Ivry, operating at 90 bars, has more breakdowns than the older plant at St. Ouen, operating at 20 bars, although this is partly offset by the reduced maintenance problems at Ivry).

Although the plants are owned by the City of Paris, they are operated by Traitement Industriel des Residues Urbains (TIRU) which is staffed by personnel from Electricité de France.

Research at the Bureau of Geologic and Mining Research (BRGM), Orleans: BRGM is a pseudo-public company which does research funded by its customers. It has been working on recovery from postincinerator waste, using a process similar to (but a little simpler than) the U.S. Bureau of Mines process. It has a pilot plant of 1 ton/hour capacity but it is hoping to construct a full-scale, 20 tons/hour facility, possibly in Paris (although TIRU is not yet convinced of its viability).

BRGM is monitoring existing work on front-end fuel/materials separation in the United States and United Kingdom. A consortium (including the Ministry of the Quality of Life and waste treatment equipment manufacturers) is funding a $1 million project that should lead to a pilot fuel/materials separation plant within 2 to 3 years.

ITALY

National Overview

General information was not obtained from Italian sources.

Notable Developments

Recovery Plants in Rome: Four companies are responsible for the disposal of about 750,000 tons/year of refuse, in the northern, eastern, southern, and western sectors of the city, respectively. The plant in the western section, receiving about 29% of the total refuse, converts it by biostabilizer into fine compost (28% of the input) and medium compost (29%); 28% is incinerated, 2% recovered as ferrous metal, and 13% is lost as moisture.

The other three plants recover paper, animal food, compost and ferrous materials, the remainder being incinerated. The most up-to-date of the plants (in the eastern sector) has three 12.5 ton/hour sorting lines and treats an average 600 tons/day (Figure 1.5). A fourth line, receiving wastes from special sources such as markets, street sweeping, etc., conveys them unsorted to the composting section.

Wastes entering the three sorting lines pass through a device that tears the plastic bags in which they are contained, and they are then jet-steamed to

FIGURE 1.5: ROME SOLID WASTE PLANT (EASTERN SECTOR)

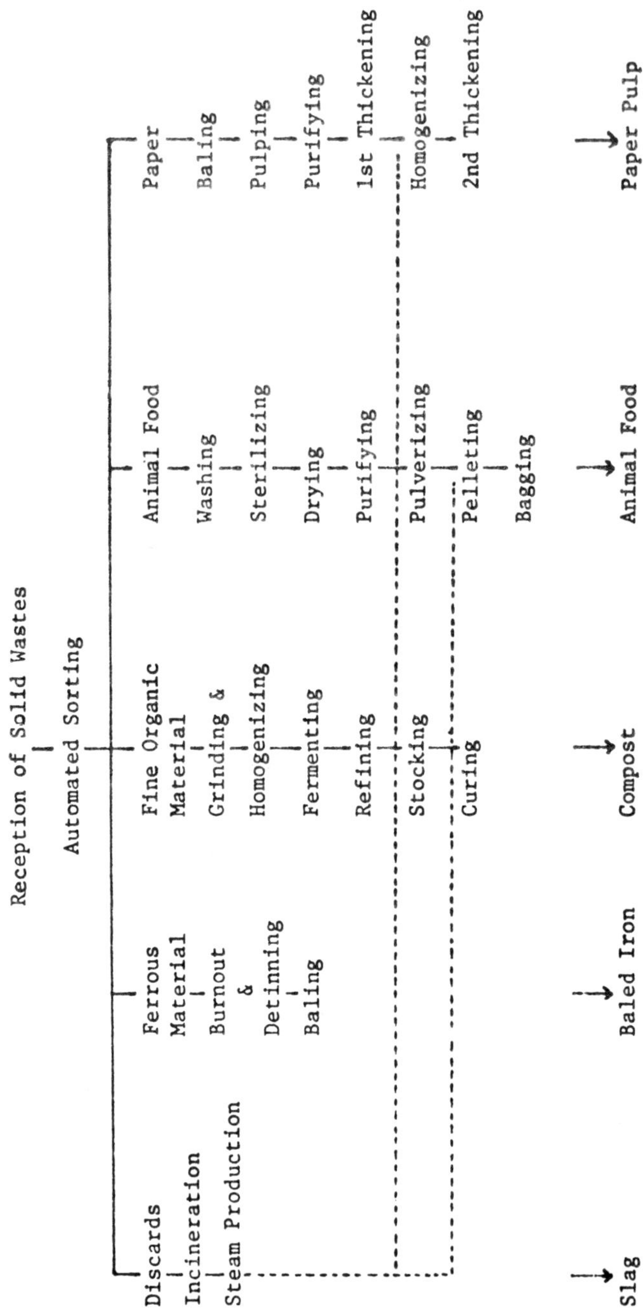

hinder the decay process. The wastes then enter a series of screens to separate (1) bulky refuse or discards which, after magnetic separation of ferrous materials, are conveyed to the incinerators; (2) medium-bulky material, from which ferrous items are extracted magnetically and paper is separated by means of an air stream is directed tangentially at the conveyor, the residue going to the composting section; (3) medium-size material which undergoes magnetic separation before being conveyed to the animal food section; and (4) fine material which goes directly to the composting section.

The section handling the recovered ferrous material includes a rotary furnace fired by liquid fuel in which the metal is detinned, cleaned, and oxidized. Material going to the paper section is baled and then pulped; foreign matter is removed by screening, etc., and the pulp is passed via a prethickener and dewatering press to a grinding-homogenizing machine where it is finally treated with saturated steam to separate and remove paraffin and tar residues as well as to open the fibers.

In the animal food section, the vegetable material is first separated from the remainder, the latter being conveyed to the composting section or to the incinerators. The separation is performed by a special hydraulic-mechanical system which also washes and rinses the material and precipitates the inerts. The vegetable material is homogenized and sterilized in autoclaves using saturated steam produced by the incinerators. The product is dried to 10% moisture, further purified in pneumatic clyclone collectors, ground finely, and then pressed to facilitate packing. The food is sterile and can be preserved for several months.

Material entering the composting section is subjected to a fermenting process under controlled conditions, to a screening process, and finally to a curing process in a compost stabilization area. Other products may be added to provide enrichment.

The incineration and steam generation section consists of three furnaces producing steam which is used in the various sections of the overall plant (especially the animal food section); some steam is also conveyed to a nearby industry. Water used in the plant is treated and recycled.

Tests are now being carried out on the further recovery of plastics and glass; the former causes problems particularly in the paper and incineration sections, while the latter causes difficulties in the composting section. Glass recovery will involve the use of an optical separator.

Estimated costs range from 7,300 to 8,100 lire/ton for amortization plus operation, less 3,400 to 3,050 lire/ton proceeds from sales of materials (42% paper pulp, 41% animal food, 12% ferrous metals, and 5% compost), making net costs, 3,900 to 5,050 lire/ton. These compare favorably with an estimated 3,374 lire/ton for "controlled dumping" in Milan, and 10,450 lire/ton for incineration with electricity generation, also in Milan.

LUXEMBOURG

National Overview

No information was obtained from sources in Luxembourg, but contacts elsewhere mentioned that an incineration plant at Leudelingen, formerly planned to have three conventional 8 ton/day furnaces, would have one of these furnaces replaced by a high temperature Torrax system. Start-up was due in 1976.

NETHERLANDS

National Overview

The Stricting Verwijdering Afvalstoffen (Institute for Waste Disposal) has been established as a government institute for the purpose of advising both the government and the local authorities about waste disposal problems and energy/materials recovery. It is supported largely by the Ministry of Environment and Public Health, although about 10% of its budget comes from sponsored projects.

Some 30% of municipal waste is currently incinerated, 17% deposited after composting, 47% dumped (uncontrolled) and 6% placed in a sanitary landfill. The most important company in the composting field is N.V. Vuil Afvoer Maatschappij Amsterdam.

The central government provides no money for the construction or operation of disposal/recovery facilities, thus recovery can be carried out only if it is economical to do so. There is some research on recovery being undertaken at private companies and at universities, etc., such as the Institute for Applied Research and Technology (TNO).

Priorities are much the same as in the United States, but pyrolysis is not considered ready for implementation at the present time. There are no pilot pyrolysis plants operating in the Netherlands, but two bench-scale fluidized bed processes are being set up at TNO (160 kg/hour) and the University of Enthoven (10 kg/hour), respectively.

A paper study is being conducted on fuel separation from refuse. A 1 ton per hour line separating paper and plastics has been set up at TNO but the recovered paper is of poor quality compared with paper which has been recycled directly, and the economics are questionable. The University of Enthoven is also experimenting with plastics separation and air classification. Some other research is being carried out on tin can separation and detinning.

Notable Development

Incineration with Heat Recovery: There are four heat recovery incinerators in the Netherlands, each generating electricity. The largest plant (at Rotterdam-Botlek) has a 20 ton/hour capacity, and exhaust steam from the turbines is used to heat the brine steam in a multi-stage, flash-type desalination plant (capacity 3 × 450 m^3/hour). The iron from the slag, and also the slag itself from most of the incinerators, is occasionally reused (although markets are difficult to find).

There have been boiler corrosion problems in the three incinerators predating the Rotterdam plant. The latter operates at a relatively low temperature (360°C) and pressure (26 kg/sq cm) and has experienced no corrosion problems since 1972. However, due to the present adverse economic situation, growth in the Rotterdam area has been less than expected; thus the incinerator is operating below capacity and produces insufficient steam for the full-scale operation of the desalination facilities. In addition, two small rotary kilns designed to burn hazardous wastes have had problems in meeting air pollution standards, as well as other operating difficulties.

It is thought that new incinerators will probably be constructed without electricity generation because the capital costs are so high and there is insufficient return from the electricity produced. The prospects are better for the recovery of steam for industrial use.

SPAIN

National Overview

No general information was obtained from Spanish sources, although contacts elsewhere suggested that apart from the recent construction of a 1,000 tons/day incinerator with heat recovery in Barcelona, the only significant development is taking place at the Empresa National Adaro.

Notable Development

Research Program at the Empresa National Adaro: This private laboratory is doing research on the front-end separation of municipal solid waste. The process used is based on the system developed at the U.S. Bureau of Mines, with modifications to allow for differences in the refuse composition, etc. (e.g., a much higher proportion, up to 60% of fermentable organic material). A first-generation pilot plant was completed in 1974, and a second-generation plant was expected to start operations in September, 1975.

SWEDEN

National Overview

The government agencies principally concerned with energy/materials recovery from solid waste are (1) the Ministry of Agriculture (particularly the National Environment Protection Board), (2) the Ministry of Health and Social Affairs, (3) the Ministry of Industry (particularly the National Board for Technical Development), and (4) the 24 county administrative boards (nature conservancy sections).

Solid waste disposal is covered by tough environmental controls (e.g., regulations governing consent procedures for waste treatment plants, the prevention of pollution, etc.). The government strongly encourages energy/materials recovery and provides financial support for construction (e.g., up to 50% of the cost of recovery plant) as well as supporting research and development.

No particular system is favored at this time; several are being investigated, including composting (e.g., one to test the composting of municipal refuse with sewage sludge), pyrolysis, fuel separation, etc. The population density is fairly low outside the cities, and the predominant method of disposal currently is landfill. Stockholm has two incinerators: one is old and has no heat recovery other than provision for some drying of sewage sludge; the other is new and generates electricity. Gothenburg has a new incinerator (owned by a special company formed by the towns in the region); the plant handles industrial as well as municipal waste and provides heat for a district heating system. There are several other incinerator plants in Sweden, but there are no plans to build more in the foreseeable future.

Separation of paper by households and separate collection by communities will be mandatory within five years (other materials may follow). The two largest paper plants are currently installing the necessary recycling (deinking) equipment, and a third is to follow, so that all of the paper can be absorbed.

Experiments are now in progress in various parts of Sweden involving the continuous separation of refuse in about 130,000 households. The materials separated are dry paper (i.e., newspapers, magazines, corrugated cardboard, cartons, bags), glass (bottles and jars), and sheet metal (tins). The experiments involve both multi-family dwellings and detached houses and are being conducted under the auspices of the ASAB foundation (including representatives from the scrap trade, the packaging industry, the Swedish Association of Local Authorities, and other interests). Random checks have shown that there is very little usable paper and glass left in the refuse sacks.

There is a government tax on the sale of beverage containers, and high deposits have been introduced on returnables (though not government required). So far, while cans seem to have retained their share of the market, there has been a shift from nonreturnable to returnable bottles.

Notable Developments

Fuel/Materials Separation System at the Högdalen Incinerator in Stockholm: In collaboration with the National Board for Technical Development, the Fläkt

Company is installing a 5 tons/hour research facility adjacent to the reception hall of the incinerator. The overall cost of the project is estimated at approximately 4 million Swedish Kroner (about $900,000) of which the National Board will contribute about one-third.

A mainly dry process is being used (Figure 1.6). The refuse will pass through a primary shredder and trommel screen into an air classifier. The light fraction, consisting of about 65% of the original input, will pass via a cyclone into a secondary shredder. Heat will be used to contract the plastics content (primarily thermoplastics) and thus be readily separated from the cellulosic fibers in a second air classifier (the plastics will be more dense and will drop). Meanwhile the heavy fraction from the first air classifier will undergo magnetic separation of the ferrous material, followed by screening, froth flotation, and further air classification to separate ceramics, glass, metals, etc.

Development of the Motala Pyrogas Pyrolysis System: The Motala Company had conducted extensive pilot trials involving the pyrolysis of municipal waste at a lime-works where gas generators (of the kind used to generate coal gas) are still in use. Based on these trials they designed the Pyrogas system (Figure 1.7), which can be used to gasify various solid fuels including municipal and industrial wastes, peat, sludge, etc. The generator comprises two stages with an internal cross-section of 10 square meters. The fuel passes through various zones in the generator, its temperature rising as high as 1500°C. The outputs include a gaseous fuel, a tar, and a slag. The tar can be burned in an oil-fired boiler and the slag, which is sterile, can be deposited without risk of pollution.

It is thought that a Pyrogas unit can pyrolyze at least 2 tons/hour of municipal waste. The first plant was to begin operation in autumn, 1975, at Gislaved (in southern Sweden) where 0.8 tons/hour of coal will initially be added. The process is guaranteed by the manufacturers to work with this input, but it is thought that the amount of coal can be subsequently reduced, thereby raising the capacity for refuse. The first six months' operation will represent a trial period in which it is hoped to answer remaining questions about the materials and energy flows as well as about pollution problems.

It is known that 1 ton of Swedish refuse will substitute about 0.2 m^3 of oil. With the 2 tons/hour guaranteed capacity, the Gislaved plant will give about 15,000 tons annual capacity, substituting about 300 m^3 of oil. If no coal is needed for the process, the capacity will be doubled.

The Gislaved plant will be owned by Gummifabriken Gislaved Aktiebolag (a rubber company). The National Board for Technical Development supported some of the early development, and the National Board for Environmental Protection is providing 50% of the costs of constructing the basic plant at Gislaved. Rubber waste will be included in the waste pyrolyzed, and the resulting gas will be burned in a furnace to provide heat for the factory.

FIGURE 1.6: FLÄKT'S RRR-SYSTEM (RESOURCES RECYCLING FROM REFUSE)

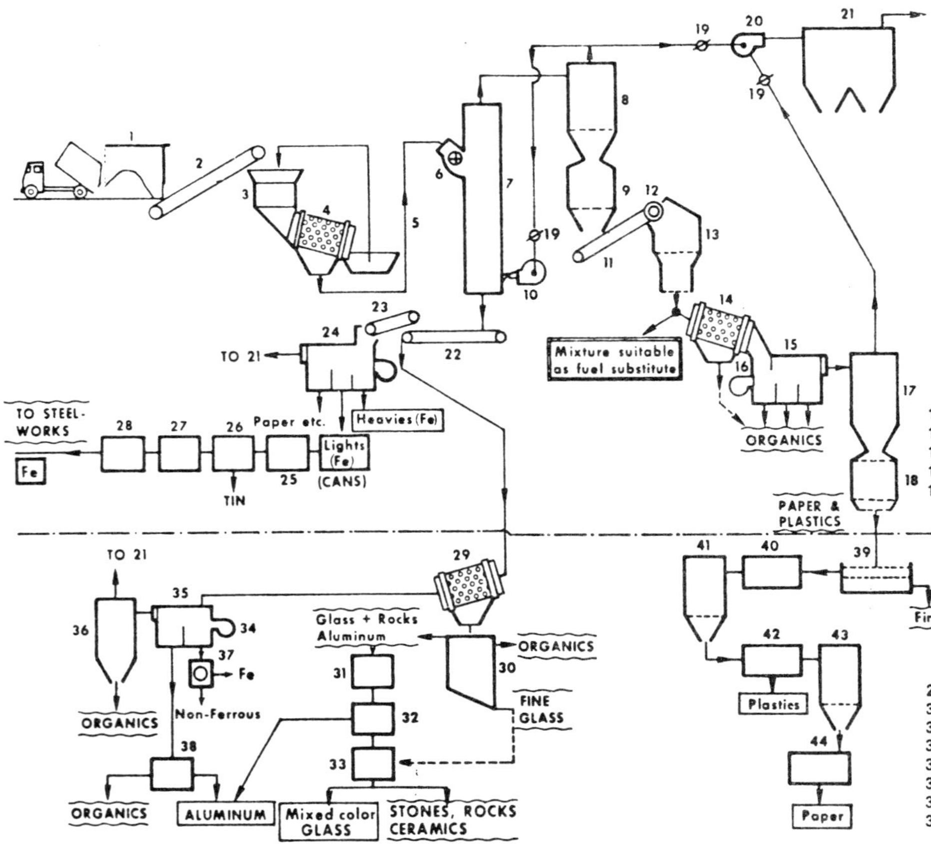

FRONT-END

1. Charging station
2. Belt conveyor
3. Primary shredder
4. Rotating trommel screen
5. Belt conveyor
6. Rotary feeder
7. Vertical air classifier
8. Cyclone
9. Container
10. Transport fan
11. Bell conveyor
12. Magnetic pulley
13. Secondary shredder
14. Rotating trommel screen
15. Horizontal air classifier
16. Transport fan
17. Cyclone
18. Container
19. Throttling valves
20. Transport fan
21. Fabric collector
22. Belt conveyor
23. Suspended electromagnet
24. Horizontal air classifier
25. Chopper
26. De-tinning
27. Surface treatment
28. Baler

BACK-END

29. Rotating trommel screen
30. Mineral jig
31. Crusher
32. Screen
33. Froth flotation
34. Fan
35. Horizontal air classifier
36. Cyclone
37. Magnetic drum
38. Electrostatic separator
39. Screen
40. Dryer
41. Cyclone
42. Air classifier
43. Cyclone
44. Paper baler

FIGURE 1.7: MOTALA PYROGAS

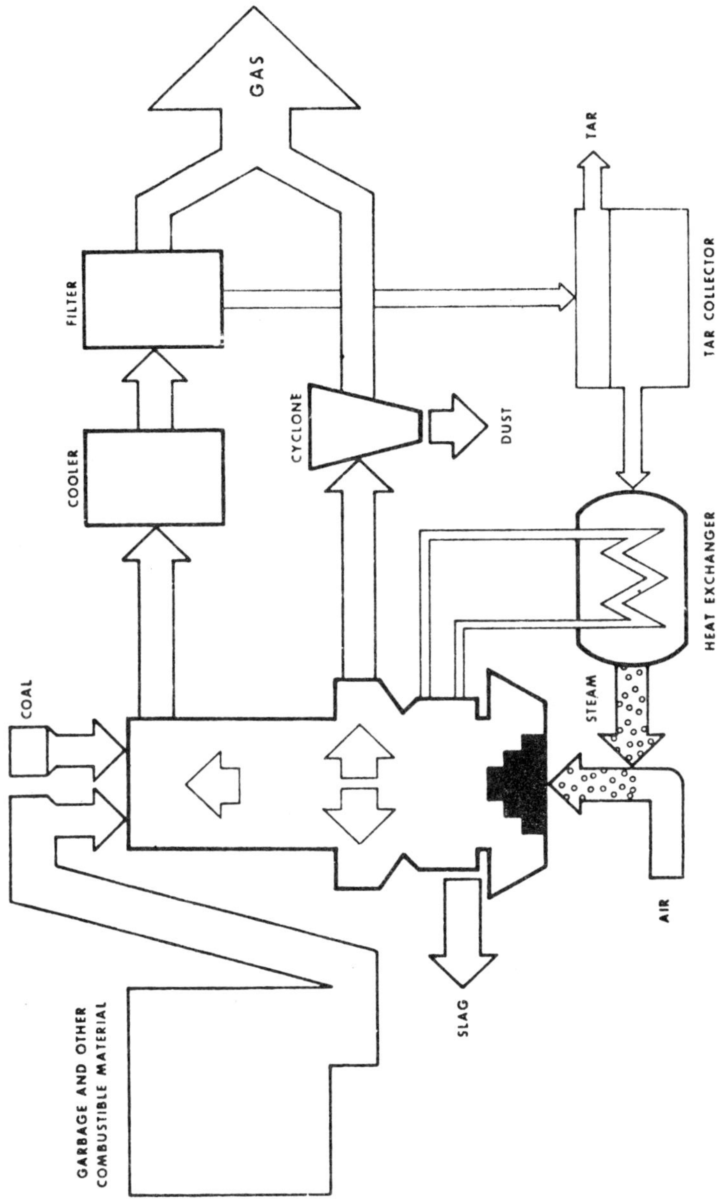

SWITZERLAND

National Overview

Although the Federal Agency for Environmental Protection supervises in a general way refuse disposal and has some influence over the choice of methods used, the local (or sometimes regional) authorities make their own decisions. The first priority has always been refuse disposal rather than energy/materials recovery (although this may be changing due to rising fuel costs).

Only the larger incinerators (say, over 200 tons/day) have heat recovery, providing electricity and/or district heating or industrial steam (the latter being preferred whenever possible as it is the easiest and most economical to operate). There are four incinerators with power generation alone (burning a total of 249,000 tons/year), eight with heat and power generation (561,000 tons per year), twenty-nine with no recovery (371,000 tons/year), and twelve composting plants (86,000 tons/year). There are also both controlled and uncontrolled landfills. It is becoming increasingly difficult to find acceptable sites for incinerators within cities or on the outskirts, and it is even more difficult to obtain land for a landfill (especially because of the mountains).

Materials recovery is performed by private firms only (except where separation is necessary to produce acceptable compost). Incinerator operators tend to oppose separate paper collections because of the reduction in the fuel value of the refuse. There is some separate glass and metals collection, although the glass industry is interested only in color-sorted glass. There is very little postincineration ferrous metal separation since few steel mills will accept the product (due to impurities). Tests are being made on the use of incinerator ash for road-building (with promising results so far) which is particularly important since Switzerland faces a gravel shortage. There has been successful laboratory experimentation on the recovery of heavy metals (e.g., zinc and lead) from fly ash.

There is a trend toward reusable bottles. Private initiative has led to the sale of certain beverages (including wines) in gauged bottles with ½ and/or 1 liter etched markings. All breweries are moving toward the use of the same bottles. Some large supermarkets are agreeing to use the same interchangeable screw-top bottles.

Notable Development

Incineration by Von Roll (headquarters in Zurich): A "typical" Von Roll plant has a capacity of around 600 to 900 tons/day (2 or 3 units of 300 tons/day), with no pretreatment of the refuse, a heat recovery system producing power and/or district heating (plus some live steam, perhaps), a wet ash/residue system, and an electrostatic precipitator. The boiler is located after the main combustion chamber and is a suspended tube panel type, with cleaning by an automatic rapping/vibrating device. The combustion chamber temperature lies between 800° to 1000°C and the gases exit into the electrostatic precipitator at 220° to 280°C. There is not normally a gas scrubber, but one is going to be necessary in the future in both Germany and Japan (due to stringent air pollution regulations).

Von Roll does not recommend viewing a refuse incinerator as a power generator. Optimal steam conditions are different for the different purposes and a compromise has to be made. Boiler corrosion is now fairly well under control but unpredictable problems can arise. Nevertheless, it is considered well established that a large incinerator can break even on its operating costs/heat recovery, and possibly a slight profit can be made (if capital costs are not included).

The materials flow for a typical incinerator includes as little as ½ ton of input water (possibly clarified sewage water) per ton of refuse burned. Each ton burned gives 350 to 450 kilos of solid residue (dry weight) with 15 to 20% water content, depending on the refuse composition. The flue gas contains a maximum of 2000 mg/nm^3 hydrogen chloride, 3000 mg/nm^3 sulfur dioxide, and traces of carbon monoxide. The carbon dioxide content is typically 5 to 11%. It is very difficult to measure heavy metals in the flue gases; it is known that most are removed at the beginning of the electrostatic precipitator.

In all new incinerators in Germany, as well as in new and certain old incinerators (over a specified size) in Japan, electrostatic precipitation is to be followed by full or partial scrubbing (depending on the size of plant, emissions allowed, etc.). The process may be wet or dry; the latter involves blowing an alkaline chemical such as dolomite into the combustion chamber (e.g., the Belgian Solvay process). The gases may or may not be reheated before emission to the atmosphere. Inevitably there are significantly increased costs, a greater risk of corrosion, and the possibility of water pollution.

Combined refuse/sewage sludge treatment is possible. At a 150 tons/year plant which has been constructed in Dieppe, France, the heat from the combustion of refuse is used to dewater and dry the sludge (previously prethickened from 98% to 80% water); the dry sludge (40 to 45% water) can then be added to the refuse while maintaining self-sustained combustion. Two similar plants are planned for Brive and Deauville (also in France). However, the sludge drying unit is very expensive (it is a jacketed vertical column type heat exchanger with critical parts made of stainless steel), and the process is probably suitable only for smaller units.

In larger plants it is preferable to preheat prethickened sludge for about 20 minutes in a continuous operation, raising its temperature from 20° to 100°C (thermal conditioning). This changes the coagulation properties, making the sludge easier to settle. The thickened material from the bottom of the vessel is dewatered in a filter press to about 40 to 45% water; it is then a sterile, odorless cake which is safe to deposit and can be used as a fuel (if ground to reduce its size) or as a soil improver, etc.

A major problem with combined refuse/sewage sludge disposal is that the authorities responsible for refuse and sludge are often separate; thus there are political rather than technical difficulties.

Von Roll considers that the burn-out in the supplementary fuel plant at St. Louis is poor. This is not a major problem where a landfill is available but it makes the use of the ash for road construction, etc., more difficult. Von Roll considers that if 10% refuse is burned, the fuel with the best burning characteristics (i.e., the normal fuel) will burn best leaving the poorer fuel (i.e., the refuse) making up most of the unburned material in the ash. Possibly an afterburning grate might be needed.

WEST GERMANY

National Overview

The government agencies principally concerned with energy/materials recovery from municipal solid waste are (1) Umweltbundesamt (UBA) (Federal Environmental Protection Agency) as well as federal Ministries such as Interior, Science and Technology, etc., and (2) state agencies, e.g., Bayerisches Landesamt fur Umweltschutz (Bavarian State Agency for Environmental Protection).

With the help of experts from industry, the universities, other government departments, etc., the UBA has prepared a major solid waste program which was expected to be adopted by the government and issued toward the end of 1975. Solid waste management is currently covered by the Waste Disposal Law of June, 1972, which includes provisions for regional planning, control of solid waste facilities, limitations on the introduction of single-use packaging, etc.

Disposal in West Germany currently involves landfill, composting, and incineration. The government is studying pyrolysis, hydrolysis, and high-temperature incineration, and is developing a major project on energy/materials recovery (the "Reutlingen" project, see below). While it is thought that air pollution problems from pyrolysis plants might be less severe than those from incinerators, there are instead problems of water pollution and tars, which are being examined. Also being examined is the problem of removing chlorides from scrubber water (for the time being this water is lagooned or fed through a municipal sewage treatment plant). These and other research and demonstration projects will be supported in 1976 by the Ministry of the Interior (from a fund of 4 to 6 million D-Mark), with UBA providing administration and technical assistance. There may be additional funds from other ministries.

There are presently some 19 composting facilities in West Germany, and the percentage of waste disposed of by composting seems to be rising from about 2% to perhaps 5% (e.g., the plan for Stuttgart calls for an increase in the number of plants from the present 6 to a total of 19). There may be a trend away from using fully mechanized plants because of high costs: new plants are likely to involve shredding, mixing with other materials (e.g., sewage sludge) and placing in large windrows with mechanical mixing. However, mechanical mixing equipment for large plants is currently not very good, although development work is proceeding in East Germany and France. Markets are not strong in all areas, but there is a constant demand from vineyard owners.

There are currently 31 incineration plants, of which 27 recover energy. Several more plants are nearing completion. Some produce electricity but the problems of doing so (i.e., corrosion, difficulties of matching supply and demand, and questionable economics) have made the production of steam for other purposes (e.g., district heating, industrial uses, distilled water production, sludge drying) more attractive. A major new consideration is the need to meet very stringent air pollution standards, necessitating scrubbing of at least part of the gas steam in addition to electrostatic precipitation (five new plants are to have both types of cleaning). This significantly raises the costs (by about 25%). Nevertheless, incinerator technology is very advanced, and it is thought that large incinerators will continue to play a role in the disposal of very large quantities of refuse.

An interesting practice is the conversion of old coal-burning power or industrial furnaces into refuse incinerators, making use of the existing infrastructure. The refuse may be burned alone or in combination with pulverized coal or with sewage sludge. An additional advantage is the fact that new planning permission is not required.

Notable Developments

The "Reutlingen" Recycling Demonstration Plant: Feasibility studies are presently being completed for the construction of a demonstration resource recovery plant probably to be located about 30 miles south of Stuttgart in a rural area, adjacent to a new composting plant, and serving about ½ million people. The process would initially involve shredding and air classification, with the light fraction providing paper recovery and the heavy fraction providing both aggregate material and composting feedstock. Market studies so far suggest that only paper is worth recovering, although it is anticipated that plastics might subsequently be recovered, followed by nonferrous metals; glass recovery is not yet considered feasible. The purpose of the project is to test out the markets as well as the processes.

It is expected that there will be three lines each with a capacity of about 15 tons/hour; different processes will be tried out, leading ultimately to the choice of an optimum configuration. Wet processes will be avoided as far as possible in order to avoid water pollution problems.

The plant will probably be owned by an independent company formed by the federal government, the states, and private industry (both the manufacturers of disposal equipment and the users of the recovered products). It is anticipated that the federal government might provide up to 40 to 50% of the costs, the states about 10%, and the industry the remainder. It should be noted that these estimates are tentative only.

Development at Krauss-Maffei (Munich): The Krauss-Maffei Company, which will be assisting in the "Reutlingen" project, has developed a front-end separation process (*System R80*), currently at pilot-plant stage. The process includes primary magnetic extraction, shredding, secondary magnetic extraction, screening and air classification (Figure 1.8). The air classifier separates the waste into three fractions, (1) the light fraction consisting of paper, plastic film, and light textiles, (2) the middle fraction consisting of cardboard, heavier plastics, leather wastes, etc., and (3) the heavy fraction consisting of the heaviest plastics, nonferrous metals, bones, wood, stones, etc.

FIGURE 1.8: KRAUSS-MAFFEI (SYSTEM R80)

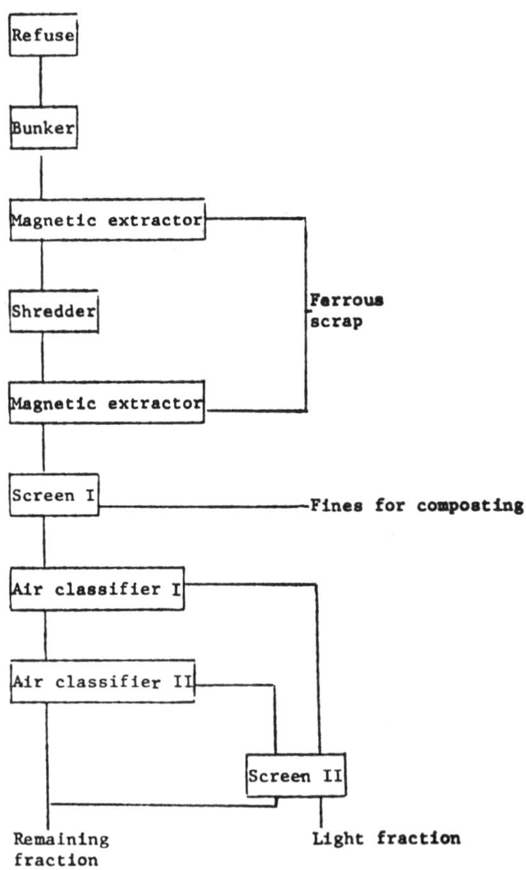

Additional components can be added to the process depending on the particular conditions, the products required, etc. The products can include a composting feedstock, recovered papers and cartons (which can be reprocessed to form new paper and cardboard), plastic granules, glass cullet, aggregate, and/or a combustible fuel (which can be incinerated or pyrolyzed).

Development at the University of Hamburg: Research on pyrolysis is reportedly being done at this university involving a bench scale fluidized bed system. No further information was obtained.

Development at Kiener, Goldshofe: A system has been developed which involves the gasification of coarsely shredded refuse in a rotary kiln. The gas produced is cracked in a gas generator and then burned in a gas engine to produce electric power. A pilot plant of 40 kg/hour capacity was originally constructed; however, this has now been shut down and a 100 kg/hour demonstration plant was started up in Goldshofe (August, 1975). The University of Stuttgart has been testing the system.

LIST OF ORGANIZATIONS AND INDIVIDUALS CONTACTED

Belgium

Antwerp Waste Management Service
St. Andriesplaats 25., Antwerp.
ir. H. Herreman, Director
M. Flament, Deputy Director

Denmark

Pollution Control Ltd.
Frydenlundsvej 30, DK - 2950 Vedbaek.
Ole H. Willerup, Managing Director

England

Department of the Environment
2 Marsham Street, London SW1P 3EB
J. Sumner, Director, Wastes Division

Department of Industry, Warren Spring Laboratory
Gunnels Wood Road, Stevenage, Herts. SG1 2BS
W.S. Douglas, Materials Recovery Division
Maurice Webb, Materials Recovery Division
Robert Flaim, Materials Handling Division

Greater London Council, Department of Public Health Engineering
10 Great George Street, London SW1P 3AB
D. Ayres, Director
P.K. Patrick, General Manager, Solid Wastes
A. Marchant
J. Ottley

H. Brown, Energy and Resources Consultant
10 Westlake Drive, Hayes, Bromley, Kent BR2 7HF

Associated Portland Cement Manufacturers, Commercial Development Div.
Aspdin House, Crete Hall Road, Northfleet, Kent
P. Giles, Project Consultant

West Midlands County Council, County Waste Disposal Department
County Hall, 1 Lancaster Circus, Queensway, Birmingham B4 7DJ
K. Harvey, Waste Disposal Officer
L. Drury

Imperial Metal Industries (Kynoch) Ltd.
P.O. Box 216, Kynoch Works, Birmingham B6 7BA
J.E. Marshall, Engineering Director, Witton Site Services Division

Staffordshire County Council
4 Chapel Street, Stafford, ST16 2BX
J. Skitt, County Waste Disposal Engineer

Open University, Faculty of Technology
Walton Hall, Milton Keynes MK7 6AA
Dr. A. Porteus

Environmental Resources Ltd.
35a Thayer Street, London W1M 5LH
Florence Fisher, Managing Director

Institute of Fuel
18 Devonshire Street, Portland Place, London W1N 2AU
Dr. R. Jackson, Secretary

France

Ministère de la Qualité de la Vie, Services des Problèmes de Déchets
14 avenue du General Leclerc, 92200 Neuilly
M. Chambournier

Direction Generale de l'Amenagement Urbain
98 quai de la Rapée, 75570 Paris Cedex 12
R. Dorfmann, Vice-Président Délégué à l'I.S.W.A.

B.R.G.M. Orleans
Jean Noël Gony

Electricité de France, Traitement Industriel des Résidus Urbains
134 boulevard Haussmann, 75008 Paris
Pierre Fourment, Chef du Service des Relations Commerciales

Italy

U.S. Embassy, Rome
John B.L. Manniello, Scientific Attache

Netherlands

Stichting Verwijdering Afvalstoffen
Amersfoort, Postbus 184, Natriumweg 7
Ir. J.A. van der Kuil, Head of Economic/Legal Department; Mr. Lueikhuize

Spain

Empresa National Adaro
Serrano 116, Madrid 6
Dr. M. Cavanna

Sweden

Statens Naturvårdsverk
Fack, S-171 20 Solna 1
S.O. Hellgren

A.B. Motala Verkstad
Fack, S-591 01 Motala
John Roslund

Goteborgsregionens Avfallsaktiebolag
Box 341, 401 25 Göteborg 1
Bengt Rundqwist

Stockholms Energiverk
Tulegatan 7-13, Fack, Stockholm 19
Ing. Tore Tulin

Switzerland

Von Roll AG
CH-8021 Zürich, Uraniastr. 31/33
C.D. Herrmann
Cornelius Solotemaker
Mr. Nuesch

Federal Institute of Technology, Department of Mechanical Engineering
CH-8006 Zürich, Sonneggstrasse 3
Professor Dr. Fritz Widmer

Federal Institute for Water Resources and Water Pollution Control (EAWAG)
CH-8600 Dübendorf-Zürich
K.A. Wuhrmann

West Germany

Umweltbundesamt
D1 Berlin 33, Bismarckplatz 1
Dipl. Ing. Werner Schenkel, Erster Direktor und Professor
Manuela Jacobs
Hans Langer

Krauss-Maffei
8000 München-54, Tannenweg 4
Dr. Hillekamp

Universität Stuttgart, Institut für Siedlungswasserbau und Wassergütewirtschaft, 7 Stuttgart 80 (Büsnau), Bandtale 1
o. Prof. Dr. Ing. Oktay Tabasaran

Bayerische Landesamt für Umweltshultz
8 München 81, Rosenkavalierplatz 3
Bartholomaeus Furmaier, Baudirektor

United States (Preliminary contacts and review of report)

First International Conference on Conversion of Refuse to Energy (CRE)
Suite 500, 299 Park Avenue, New York, New York 10017
W.K. MacAdam, Conference Vice-Chairman

American Public Works Association
1313 East 60th Street, Chicago, Illinois 60637
Rodney R. Fleming, Associate Executive Director

Solid Wastes Management-Refuse Removal Journal
461 Eighth Avenue, New York, New York 10001
Eugene L. Pollock, Editor

National Center for Resource Recovery, Inc.
1211 Connecticut Avenue, N.W., Washington, D.C. 20036
Dr. Harvey Alter, Director of Research Programs

REFERENCES

England

Douglas, E., Power, C., and Walsh, T., "Development of a Flowsheet for Extracting Non-Ferrous Metals from Domestic Refuse Clinker," *Proceedings of International Symposium on Solid Waste Disposal*, Montreal, September 1974.

Douglas, E., Webb, M., and Daborn, G.R., "The Pyrolysis of Waste and Product Assessment," *Proceedings of Symposium on the Treatment and Recycling of Solid Wastes,* Institute of Solid Waste Management, Manchester, January 1974.

Douglas, E. and Birch, P.R., "Recovery of Potentially Re-Usable Materials From Domestic Refuse by Physical Sorting," Prepared for Symposium on The Technology of Reclamation, University of Birmingham, April 1975.

Patrick, P.K., "Operational Experience in Energy Recovery Through Incineration," Presented at International Symposium on Energy Recovery From Refuse, University of Louisville, September 1975.

Porteus, A., "The Recovery of Fermentation Products from Cellulose Wastes via Acid Hydrolysis," Open University, February 1975.

Sumner, J., "Waste Disposal—A New Philosophy," Presented at 101st Annual Conference of the Institution of Municipal Engineers, Torbay, June 1974.

France

"Les Déchets Solides—Propositions Pour une Politique," Rapport du Groupe d'Études sur l'Élimination des Résidus Solides, La Documentation Française, Paris 1974.

Italy

Frangipane, E. de Fraja, and G. Bozzini, "Municipal Solid Waste Disposal by Recovery Plants in Rome," Translated by American Embassy, Rome, August 1975.

Sweden

Hovenius, G., "Composting Project at Laxå—Examination of Tenders of Mechanical Equipment," Statens Naturvårdsverk, 1975.

Bahri, M., "Resources Recovery from Municipal Solid Waste—Full Scale Tests," Flakt (AB Svenska Flaktfabriken), December 1974.

"Regional Development Plan Serves Large Community," *Solid Wastes Management, 17,* December 1974.

National Swedish Environment Protection Board, "Solid Waste Management in Sweden," Prepared for ECE Seminar on the *Collection, Disposal, Treatment and Recycling of Solid Wastes,* Hamburg, September 1975.

National Swedish Board for Technical Development, "Pyrolysis of Household Waste," May 1974.

West Germany

Schenkel, W., "Technological and Economic Feasibility, Research and Policy Principles Related to Collection, Transportation and Disposal of Solid Wastes," Prepared for ECE Seminar on the *Collection, Disposal, Treatment and Recycling of Solid Wastes,* Hamburg, September 1975.

Umweltbundesamt (Author: Klaus Stief), "National Information About the General Situation and Main Problem Areas in the Field of Solid Waste Management," Prepared for the ECE Seminar on the *Collection, Disposal, Treatment and Recycling of Solid Wastes,* Hamburg, September 1975.

Case Study of Toulouse-Le Mirail, France

Material for this chapter has been based upon a report prepared by Resource Planning Associates, Inc., Washington, D.C. and Paris (CONS/2103-1).

SUMMARY

The waste treatment system of Toulouse-Le Mirail (France) is a complex composed of an incinerator with heat recovery equipment and an oil-fired heating facility. It was designed to burn the household waste of a city of 400,000 inhabitants, while at the same time providing central heating and hot water to a suburb of 100,000 inhabitants.

The choice of incineration as a means of waste treatment by the city of Toulouse was originally due to the lack of suitable landfill sites in the area. The decision to recover the heat was closely linked to the beginning of the development of a complete new suburb of 100,000 inhabitants. This new suburb had considerable energy requirements for central heating and hot water, and a district heating network was planned. Moreover, the municipality of Toulouse was receptive to the idea of heat recovery from waste, as an old incinerator with heat recovery had been operating since 1926.

Construction of the thermal complex was carried out in two phases: the first two incinerators and the oil-fired heating facility began operation in 1970; the third incinerator was completed in 1975.

The three furnaces (see Table 2.1) are equipped with roller grate stokers of the VKW Dusseldorf type; each stoker is composed of six steel rollers containing cast iron inserts; the air for combustion is blown through the inside of the rol-

TABLE 2.1: TOULOUSE-LE MIRAIL THERMAL COMPLEX TECHNICAL DATA

Dates		
Start of construction	March 1968	
First incineration of household waste	October 1969	
Completion of Phase I	January 1970	
Completion of Phase II	August 1975	
Plant Dimensions	**Metric System**	**English Equivalent**
Area of site	4.5 hectares	11 acres
Area of building	5,700 m^2	61,000 sq ft
Dimensions of building	119 m x 54 m	390 ft x 177 ft
Height of central block	30 m	98 ft
Height of smokestack	45 m	148 ft
–diameter of first flue	4.4 m	14.4 ft
–diameter of second flue	2.4 m	7.9 ft
Capacity of Transformers	6,000 KVA	
Weight of the Structural Steel	1,200 metric tons	1,320 short tons
Incinerators		
Number of furnaces	3	
Rated capacity (per hour, per furnace)	8.15 metric tons	8.97 short tons
Minimal capacity (per hour, per furnace)	4.8 metric tons	5.3 short tons
Steam production per furnace at normal capacity, lower heating value of 2,200 kcal/kg (1,000 Btu/lb) of waste	23.2 metric tons/hour	25.5 short tons/hour
Height of furnaces	24 m	79 ft
Volume of waste pit	4,300 m^3	5,624 cu yds
Volume of crane bucket	3,600 litres	4.7 cu yds
Volume of residue pit	630 m^3	824 cu yds
Flue gas produced (per hour, per furnace)	46,500 m^3	1,642,000 cu ft
Efficiency of electrostatic precipitator	97%	
Maximum quantity of particulates	150 mg/m^3	
Average temperature of gases	235° C	455° F
Daily capacity per furnace	195 metric tons	215 short tons
Daily capacity for entire plant	585 metric tons	644 short tons
Heat Recovery System		
1 turbo-generator (1,500 kW)	16 bars/2 bars	15.8 atm/1.9 atm
1 heat exchanger for hot water	2 kth/hour	7.94×10^6 Btu/hour
1 low-pressure heat exchanger for heating	16 kth/hour	6.35×10^7 Btu/hour
1 high-pressure heat exchanger for heating	10 kth/hour	3.97×10^7 Btu/hour
Heating Facility		
1 generator	20 kth/hour	7.94×10^7 Btu/hour
1 generator	40 kth/hour	1.59×10^8 Btu/hour
Maximum temperature	210° C	410° F
Length of distribution network	2 @ 18 km	2 @ 11 mi
Pumps for distribution network	2 @ 700 hp	
Direct production of hot water for in-house use	2 kth/hour	
Capacity of water softening equipment	160 m^3/hour	42,272 gal/hour

SOURCE: BETURE

lers. Each furnace has an air preheater and a supplementary burner. The first two furnaces have classic water tube boilers. The third is a waterwall.

Each furnace has a rated capacity of 8.15 metric tons (8.97 short tons) per hour. Normal use is 24 hours a day, 330 days a year. The theoretical steam production is 23.2 metric tons (25.6 short tons) per hour with an assumed lower heating value of 2,200 kcal/kg (3970 Btu/lb).

In reality, the amount of waste treated annually is approximately 100,000 metric tons (110,000 short tons), compared with a rated capacity of 130,000 metric tons (143,000 short tons) for two furnaces and 190,000 metric tons (209,000 short tons) with the addition of the third furnace. This limitation is due to the quantity collected. The furnaces are run at a capacity of 9 to 10 metric tons (10 to 11 short tons) per hour, which is above the rated capacity of 8.15 metric tons (8.97 short tons) per hour. In fact, the furnaces are larger than needed because the lower heating value of the waste was assumed to be 2,200 kcal/kg (3970 Btu/lb), while the actual lower heating value is about 1,560 kcal/kg (2808 Btu/lb).

The energy recovered is used to provide central heating and hot water for the suburb of Le Mirail and to produce electricity for use in the plant. The electricity generated from heat recovery amounts to 5,000,000 kWh/yr; the heat recovered for central heating to 44,000 kth/yr (1.75×10^{11} Btu/yr) and the heat recovered for domestic hot water to 7,000 kth/yr (2.8×10^{10} Btu/yr), which gives a total energy recovered of 56,000 kth/yr (2.22×10^{11} Btu/yr). (One kilotherm equals 1,000 kcal or 3.97×10^6 Btu.) The thermal efficiency of the system is estimated at 65%. Twenty-five percent of the heat theoretically available is lost in the furnaces and boilers, and another 10% in the heat exchangers. Moreover, some potential heat is not recovered because of the lack of demand during the summer. Actual recovery of heat usually varies between 30% and 90% of the theoretical limit, and can even reach 96% according to the months considered. Overall, based on the experience of the 1974/75 operating year (September 1974 to August 1975), it is estimated that 40% of the energy content of the waste is effectively recovered.

Operation of the incinerator and the heating facility has been normal during the first five years of existence. No unexpected major problem has occurred. Each furnace is shut down one month per year for maintenance. Nonroutine maintenance costs (including replacement parts) during the first five years of operation ran about 1% of the cost of construction per year, rising slightly after the first three years.

In order to reduce atmospheric pollution, the gases pass through an electrostatic precipitator before escaping from a stack 45 meters (148 ft) high. The efficiency of precipitation is better than 97%. Sampling of the emissions has shown a high level of HCl, about 5 g/kg (0.005 lb/lb) of waste. The concentration of particulates is 100 mg/m^3. Nearby residents complain quite frequently but it is difficult to say at the present time if these complaints are legitimate.

The architecture of the complex is interesting although there is disagreement about its aesthetics. Working conditions in the plant have had to be improved, but they now appear satisfactory.

Forty-seven people are employed in the operation of the thermal complex, of whom 6 are concerned only with the steam distribution network. Of the 41 people working at the incinerators and the heating facility, 20, organized in 4 groups, are directly involved in operations. The workers, especially those who run the furnaces, are highly skilled.

The first phase, comprising two incinerators with heat recovery equipment, cost 25,200,000F (1969). The second phase, consisting of the addition of a third furnace, cost 17,700,000F (1975). Subsidies obtained from the national government amounted to 12% of total cost. The balance was funded through low interest loans (about 6% for the first two furnaces and 10% for the third). The amortization rate for the first two furnaces has been estimated at 34F/metric ton ($6.80/short ton; 1975) incinerated, based on approximately 100,000 metric tons (110,000 short tons) per year.

The complex is operated by a private company which receives subsidies from the municipality and repays the municipality for the heat recovered. In the operating year 1974/75, the municipality paid fees equal to 34F/metric ton ($6.80/short ton; 1974/75) incinerated, and recovered revenues equal to 27F/metric ton ($5.40/short ton; 1974/75) incinerated. Direct operating cost for the municipality was therefore 7F/metric ton ($1.40/short ton; 1974/75) incinerated.

The heat recovered is sold by the private operators of the thermal complex at 90F/kth ($5/$10^6$ Btu; 1974/75) (tax included), which is equal to or under market price. No market has yet been found for the residues of incineration which are dumped in a landfill.

The overall cost of incineration including amortization and operating costs, has been estimated at approximately 41F/metric ton ($8.20/short ton; 1974/75) incinerated, taking into account only the first two furnaces. The addition of the third furnace was probably a mistake, as it adds to the cost without increasing the amount of waste treated. The cost of incineration would be lowered by finding new sources of waste in nearby communities or by finding new clients for the heat recovered during the nonwinter periods.

BACKGROUND

Economic and Demographic Data

The thermal complex of Toulouse-Le Mirail is located in Toulouse, the fourth largest city in France with a population of 380,000. Situated in the southwest of France, Toulouse is a provincial city, the administrative and commercial center of a region of grain and general agriculture. Toulouse is, however, also a center for sophisticated industries and research; for example, the supersonic plane, Concorde, is a local product.

The city has an area of 120 km^2 (45 sq mi) with an average density of 3,200 persons/km^2 (8,300/sq mi). The historic center, which is centuries old, consists of three-story red buildings and very narrow streets. The suburban fringe, which grew considerably in the 1950s, is a mixture of large modern buildings and individual houses. Farther out, urbanization is dispersed and there are even some farms. Average income in Toulouse is comparable to that of other French cities, but with extremes of high and low incomes even more pronounced. Because of the increase in population in the 1950s and because of fear of disorderly growth, the municipality of Toulouse decided in 1960 to concentrate its expansion in one area, the suburb of Le Mirail. The goal was to build a whole new town of 100,000 people with its own administrative and commercial center. So far, 7,000 dwelling units have been built, housing 28,000 people. The thermal complex of Toulouse-Le Mirail has been an integral part of this development. The goal was to provide heat and hot water for a town of 100,000 inhabitants, while eliminating the household waste of a metropolis of 400,000 inhabitants.

Waste Collection and Treatment

Waste collection in Toulouse, as in many other French cities, is the direct responsibility of the technical services of the municipality. Household waste is collected six times a week, except in a few outlying areas with low density, where it is collected only three times a week.

The city is divided into 55 districts, each served by a team of four people with a packer truck. These districts are very dissimilar. In the historic center, small vehicles with a three metric ton (3.3 short tons) capacity are used; in the suburban fringe, trucks of 14 m^3 (18.3 cu yd) with a capacity of 5 to 8 metric tons (5.5 to 8.8 short tons) are used, each truck making two trips a day. In the suburbs, trucks of 16 m^3 (21 cu yd) with a capacity of 8 metric tons (8.8 short tons) are used (1).

Approximately 110,000 metric tons (121,000 short tons) of waste are collected annually from a population of almost 400,000 (including two other small municipalities besides Toulouse). This amount includes not only household waste but also some commercial waste coming mainly from the center of the city (2) (see Figure 2.1).

The collection service employs 270 people (of whom 20 are administrators) and 60 vehicles. The cost of collection in 1975 was 151F/metric ton ($31.90 per short ton; 1975) or 42F/inhabitant ($9.80/inhabitant; 1975). This amount is relatively high compared to other French cities (3).

Most of the waste collected is incinerated in the thermal complex of Toulouse-Le Mirail. Another incinerator at Toulouse-Le Ramier, which dates from

FIGURE 2.1: QUANTITY OF WASTE COLLECTED AND POPULATION GROWTH IN TOULOUSE 1936–1975

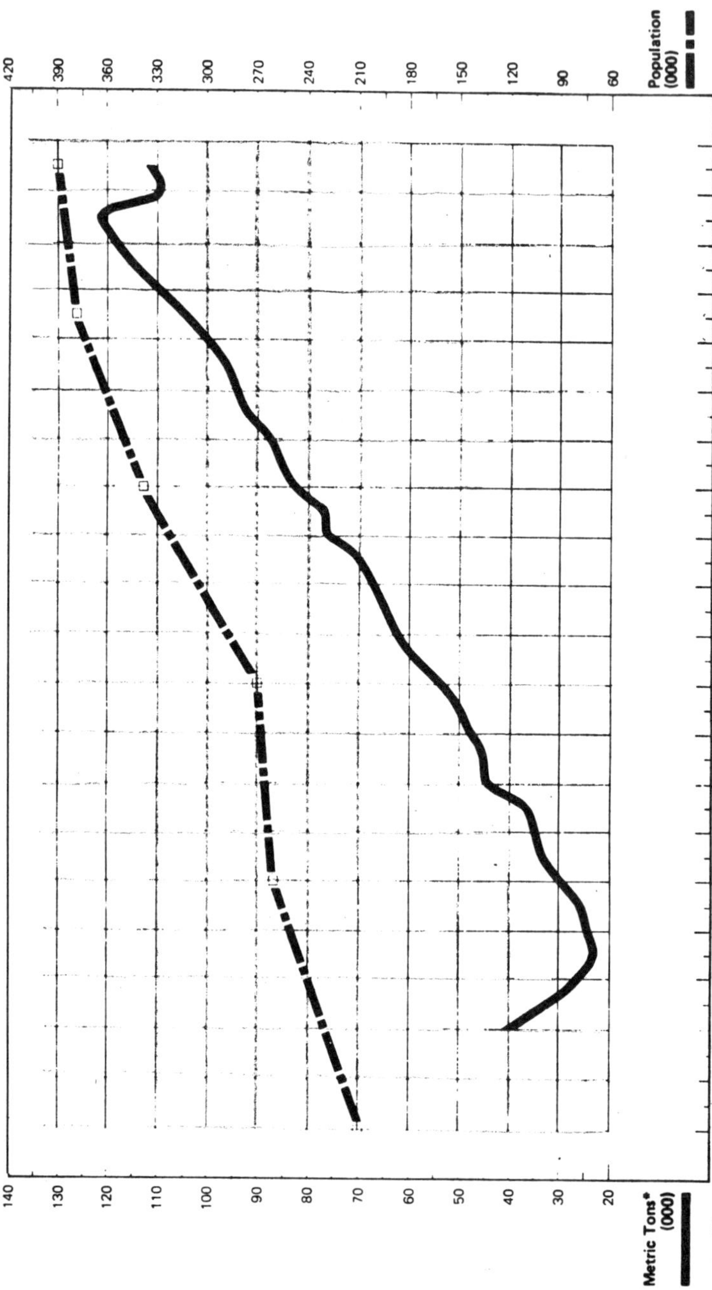

SOURCE: Noirot, "L'Usine Du Mirail–Traitement Des Ordures," Ingénieurs Des Villes De France; May, 1976.

*To convert metric tons to short tons, multiply by 1.1.

1926, treats approximately 22,000 metric tons (24,200 short tons) per year. Incinerator residues are dumped in a landfill located in a nearby community. This site is also used for other kinds of waste, such as construction and demolition waste.

DEVELOPMENT OF THE SYSTEM

Reasons for the New System

In the early 1960s, before the construction of the thermal complex of Toulouse-Le Mirail, 50% of the municipal waste was incinerated in the old incinerator of Toulouse-Le Ramier, 5% was treated in an experimental compost unit, and the remaining 45% was dumped in landfills (4).

The old incinerator of Toulouse-Le Ramier, which is equipped with Venien furnaces and heat recovery systems, had been built in 1926. After more than 35 years of operation, it was then in need of a complete overhaul. Moreover, the area in which it had been built had become one of the favorite spots in Toulouse for sports and leisure activities. Consequently, the city is planning to phase out this facility in the near future.

The Pratt compost system was tried as an experiment. This process involved aerobic fermentation in open-sided cages, with sorting and crushing carried out after fermentation. Doubts concerning the potential market for compost produced in this way and concern over competition from cities having already embarked on this experiment led Toulouse to reject further extension of the Pratt system. Given the present state of the market for compost in France, this decision seems to have been wise.

Finally, the landfills were becoming completely full. Moreover, because of the lack of suitable sites within the city limits of Toulouse, these landfills were located mainly in the nearby small town of Blagnac. In addition to the potential problems of tolerance which could be created by surrounding communities, the distance of the landfills from Toulouse (7 km or 4.3 mi) added considerably to the cost of disposal.

To all these elements, one must add the great increase in the population of Toulouse in the past twenty years and the decision, at the beginning of the 1960s, to develop the new town of Le Mirail within the Toulouse city limits, which added to the need for a new treatment system.

The final choice of incineration with heat recovery was thus linked to the development of Le Mirail. The facility which was planned to provide district heating and hot water for the dwelling units had considerable energy requirements, which could be met in part by the incinerator.

The integration of the incinerator in the development of Le Mirail was certainly facilitated by the government agency which has a financing role in such municipal operations, the *Caisse des Dépots et Consignations.* This public agency has under its corporate umbrella several engineering firms which often participate in the development of these programs. One such firm, Beture, specializes in urban and municipal engineering projects including incineration and waste treatment.

Finally, an element which was certainly important was the fact that the municipality of Toulouse had already had experience with heat recovery from incineration. The old incinerator of Toulouse-Le Ramier had been equipped with heat recovery systems and the energy recovered used to heat municipal buildings, such as the swimming pool, and to produce electricity for street lighting. This incinerator, which has operated satisfactorily for about 50 years, is still in operation in 1977.

Project Development

The preliminary studies for the incinerator with heat recovery equipment and the oil-fired heating facility were carried out by the municipality in the early 1960s. The detailed technical and economic studies which led to the specifications adopted in the invitation for bids were carried out by municipal engineers and economists.

The invitation for bids for the heating facility was published in 1965, the contract was awarded in 1966, and construction began in 1967. For the incinerator, the invitation for bids was published in 1968, construction of the first two furnaces began in 1969 and operations began in 1970. According to the contractor, a one-to-two year period is necessary for construction of an incinerator. Construction of the third furnace began early in 1974 and was completed in 1975.

SETOMIP (*Société Economique pour Toulouse et Midi Pyrennées),* a semipublic corporation (a *Société d'économie mixte,* a form of economic organization common in France) was created to manage the overall development of the new town of Le Mirail. The contracts for the heating facility and the first two incinerators were awarded by SETOMIP; the contract for the third incinerator was awarded directly by the municipality.

Invitations for bids for the incinerator and the heat recovery equipment were published in accordance with rules imposed by the national government upon municipalities for this kind of program. Bidding was open to companies selected by prequalification.

The following companies or their French representatives sent prequalification packages:

- — Martin (Germany)
- — Von Roll (Switzerland)
- — VKW (Germany)
- — Alberti Fonsard (France)
- — Combustion Engineering (U.S.)
- — Volund (Denmark)

All these companies except Alberti Fonsard actually entered bids. The selection was made on the basis of two sets of criteria.

- a complex set of governmental guidelines for comparison of such proposals (5)
- total cost.

The Tunzini engineering firm (which represents VKW in France) obtained the best rating on each of the sets of criteria (6) and received the contract. VKW, Germany participated directly in the deal, as it provided the furnaces and other equipment, as well as technical assistance.

TECHNICAL INFORMATION

The Incinerator (7)

The waste is transported by truck up a ramp to the tipping platform. The trucks are weighed upon entering and leaving, in the presence of a representative of SETMI and one from the city. The volume of the waste pit is 4,300 cubic meters (5,624 cu yd) which allows storage for two days of collection.

An overhead crane with a capacity of 3,600 liters (4.7 cu yd) mixes the waste before loading it into the hoppers. These hoppers lead directly into the combustion chambers of each furnace. A hydraulic ram at the bottom of each hopper feeds the waste into the furnace (Figure 2.2).

All three furnaces use VKW mechanical stokers. The waste from the hoppers falls onto a cascaded series of six rollers which agitate it during combustion. The rollers are made of steel with cast iron inserts. Each roller has a diameter of 1.5 m (4.9 ft) a useful length of 2.5 meters (8.2 ft) and a useful area of 32.4 m^2 (348.6 sq ft).

Air for combustion is blown through the inside of the rollers and is regulated by shutters. This air is drawn from the refuse pit, reducing the air pressure in the pit and so reducing the amount of dust and odor.

The refuse in the hopper forms a seal for the furnace. Safety of the system is ensured by a control device using radiation to monitor the waste level in the hopper. Each furnace has an air preheater and a supplementary burner for use when the waste is too damp for proper combustion. Combustion takes place at temperatures between 500° and 1000°C (932° and 1832°F), which is not excessively high.

The first two furnaces are lined with refractory brick and have classic water tube boilers. The bundles of tubes are arranged vertically. The combustion gases make four passes.

FIGURE 2.2: HEAT AND STEAM FLOW

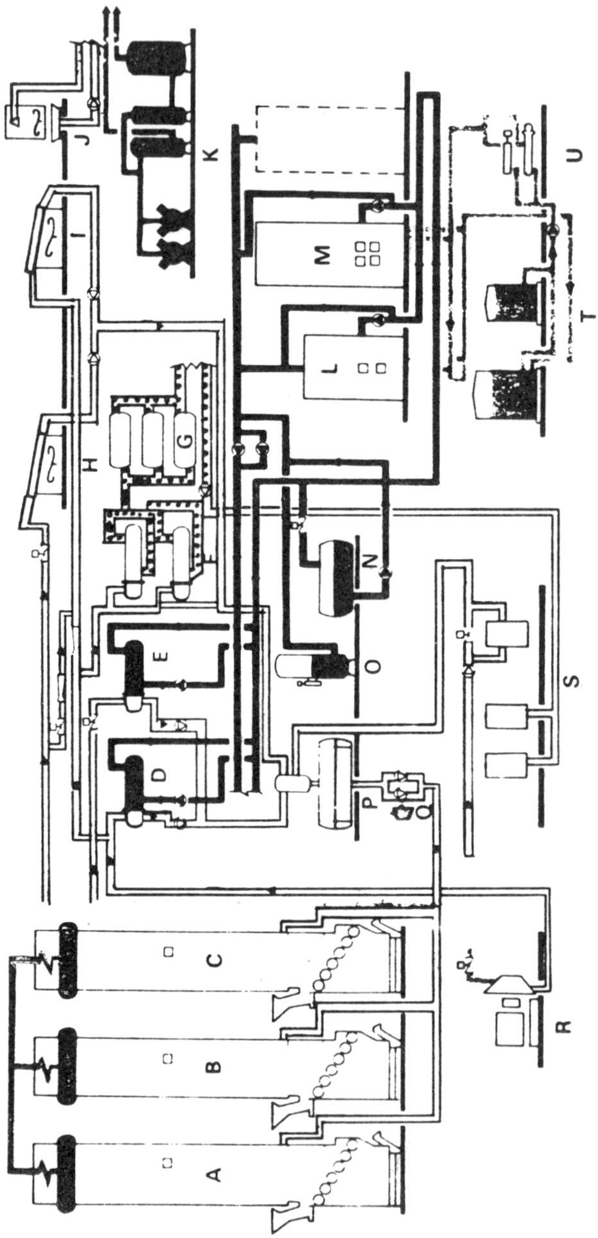

A Furnace No. 1
B Furnace No. 2
C Furnace No. 3
D Primary low pressure heat exchanger
E Primary high pressure heat exchanger
F Domestic hot water heat exchanger
G Domestic hot water tank
H High pressure condenser
I Low pressure condenser
J Cooling circuit
K Compressed air circuit
L Oil fired boiler (20 kth)
M Oil fired boiler (40 kth)
N Reservoir
O Expansion chamber
P Thermal degasser
Q Turbo-pump
R Turbo-generator
S Water softener
T Fuel oil tank
U Fuel oil preparation

The third furnace is a waterwall. The combustion chamber is lined with refractory brick to a height of approximately one meter (39 inches), so that the water tubes come down quite close to the grate. The combustion gases make only three passes in this unit.

Each of the three furnaces is 24 m (78.7 ft) high and each has an hourly rated capacity of 8.15 metric tons (8.98 short tons).

The combustion residues fall into a water-sealed ash trough. The residues from all three furnaces are gathered on a conveyor belt which carries them to a single ash pit with a capacity of 620 m^3 (810 cu yd). This pit is equipped with a grab-bucket operated from an overhead crane. The ashes are then removed by truck.

The flue gases pass through electrostatic precipitators and the dust is collected on plates which are cleaned by shaking. At the outlet of the precipitators, a fan draws the gases up into a 45 m (148 ft) smokestack.

The ash pit is also used to store the dust removed from under the roller by means of an Archimedean screw and the fly ash removed from the smoke by electrostatic precipitators.

The steam produced by incineration of domestic waste is at a pressure of 16 bars (15.8 atm). It is expanded into a turbo-generator with the following characteristics:

- intake pressure: 16 bars (15.8 atm)
- normal continuous power: 1,500 kW
- back pressure at the exhaust: 2 bars (1.97 atm)
- steam consumption: 31 metric tons (34 short tons) per hour

This system is sufficient to furnish all electric power for the thermal complex. Furthermore, the temperature of condensation of the exhaust steam, which is 134°C (273°F), allows it to be used to heat both the domestic hot water and the return water from the urban heating network.

The feed pump of each boiler is driven by a turbine also powered by steam produced by incineration. The intake pressure is 16 bars (15.8 atm) and the back pressure at the exhaust is 2 bars (1.97 atm).

The domestic hot water is heated by steam at 2 bars (1.97 atm) exhausting from the turbo-generator or from the steam turbine. The heat exchanger has a capacity of 6 kth/hour (2.38×10^7 Btu/hr).

The return water of the heating network is heated by two methods, either separately or in combination:

— back pressure steam from the turbines at 2 bars (197 atm) through a heat exchanger from 90° to 119°C (194° to 246°F)
— steam of the incineration boilers at 16 bars (15.8 atm) 10 kth/hr (3.97×10^7 Btu/hr) through a heat exchanger

The thermal complex also includes an additional recovery system for the summer period when the heating requirements of Le Mirail are very small but when the incineration must still continue. This system consists of:

— an air cooled condenser for the steam at 2 bars (1.97 atm) with a capacity of 20.5 metric tons (22.6 short tons) per hour
— an air cooled condenser for the steam at 16 bars (15.8 atm) with a capacity of 22.5 metric tons (24.7 short tons) per hour
— two heat exchangers of 6 kth (2.38×10^7 Btu) per hour between the water for heating and the domestic hot water

Heating Facility (8)

An oil-fired heating facility also provides hot water for domestic use and home heating for the new town of Le Mirail. It is located in the same complex as the incinerator (see Figure 2.2).

The thermal complex will eventually contain one boiler rated at 20 kth/hr (7.94×10^7 Btu/hr) and four boilers rated at 40 kth/hr (1.59×10^8 Btu/hr). Currently, only the boiler of 20 kth/hr and the first of the four boilers of 40 kth/hr have been installed.

In addition to the heat recovered from the incineration of waste, these boilers are also fired by oil. The heat derived from waste incineration accounts for 41% of the annual total produced by the thermal complex. In summer, almost no fuel oil is used (see Table 2.2). The boilers use heavy fuel oil which is stored in a tank with a capacity of 2,900 cubic meters (766,180 gal). A second tank is planned. The steam is superheated to 210°C (410°F) and is circulated in the heating system of Le Mirail at 30 bars (29.6 atm) by two variable speed electric pumps of 700 horsepower each.

The steam distribution network is 18 km (11 mi) long and serves the three districts of Bellefontaine, La Reynerie, and Le Mirail, or a total of 7,000 dwellings. The pipes are made of steel. A central electronic system automatically controls the steam distribution network and its 63 substations. In two of these substations, the water is reheated by natural gas. One of them is a primary school where the heating equipment already existed. The other is a school at the furthest end of the network. The other 61 substations are equipped with heat exchangers. Flow and pressure are regulated automatically.

TABLE 2.2: THERMAL BALANCE SHEET, 1974–1975

Period of Operation	Quantity of Waste Incinerated (metric tons*)	Heavy Oil Consumed (metric tons)	Heat Derived from Fuel (kth**)	Heat Recovered from Waste (kth)		Total Heat Produced (kth)	Heat Sold (kth)			Network Yield	Energy Loss in the Network (kth)
				For Home Heating	For Hot Water		Home Heating	Hot Water	Total		
September 1974	7,756	36.36	288	2,261	1,084	3,633	95	1,358	1,453	0.399	2,180
October 1974	7,721	587.80	4,651	5,902	411	10,964	6,674	1,186	7,860	0.717	3,104
November 1974	7,052	1,261.55	9,983	4,188	0	14,171	11,208	1,773	12,981	0.916	1,190
December 1974	7,724	1,539.99	12,186	3,262	92	15,540	12,553	1,861	14,414	0.927	1,126
January 1975	7,791	1,581.82	12,517	4,885	532	17,934	14,203	1,997	16,200	0.903	1,734
February 1975	6,993	1,412.46	11,177	4,679	655	16,511	12,670	1,944	14,614	0.885	1,897
March 1975	7,491	1,275.43	10,092	3,994	860	14,946	11,373	1,770	13,143	0.879	1,803
April 1975	7,909	1,175.67	9,303	4,382	951	14,636	10,700	2,116	12,816	0.875	1,820
May 1975	7,864	338.01	2,675	3,690	567	6,932	3,413	1,675	5,088	0.734	1,844
June 1975	7,602	15.50	123	2,715	868	3,706	96	1,813	1,909	0.515	1,797
July 1975	6,806	9.50	75	2,488	593	3,156	77	1,301	1,378	0.436	1,778
August 1975	6,116	6.33	50	1,379	402	1,831	63	888	951	0.519	860
Total	**88,825**	**9,240.42**	**73,120**	**43,825**	**7,015**	**123,960**	**83,125**	**19,672**	**102,807**	**0.829**	**21,133**

SOURCE: Annual Report to the City of Toulouse by BETURE, 1974-75 *(Rapport Technique et Economique).*

* To convert metric tons to short tons, multiply by 1.1.
** One kth (kilotherm) equals 3.97×10^6 Btu.

The boilers also feed a network distributing hot water for domestic use. The cold water provided by the city is first softened in an installation treating up to 160 m^3 (42,272 gal) per hour. The water is heated to 70°C (158°F) in a succession of heat exchangers using the heat provided by the incinerator. It is then stored in two horizontal insulated tanks with a capacity of 100 m^3 (26,420 gal) apiece.

The domestic hot water network has return pipes and circulates the water continuously to maintain a temperature of 65°C (149°F). The pipes are of cast iron. This network serves only the Bellefontaine district which is located about 800 m (half a mile) from the plant. Originally, there were plans to extend this service to the two other districts of Le Mirail and La Reynerie which are 3-4 km (1.8-2.4 miles) from the plant. This extension was rejected partly because of the poor quality of the water provided by this system and partly because the greater distance would make the cost of the network prohibitive. The provision of domestic hot water by this method is considered an experiment which has failed.

System Efficiency

The quantity of waste incinerated at Toulouse-Le Mirail varies between 86,000 and 100,000 metric tons (94,700 and 110,000 short tons) per year. It decreased during the years 1973 and 1974, (following a decrease in the amount of waste collected due to the economic recession) but again reached 100,000 metric tons (110,000 short tons) in 1976 (see Figure 2.1).

Theoretically, with two furnaces operating at full capacity, almost 130,000 metric tons (143,000 short tons) of waste could be incinerated annually. It is clear that, with no more than 100,000 metric tons (110,000 short tons) per year actually treated, the two furnaces have been operating below capacity.

The addition of a third furnace at the end of 1975 has increased this problem by bringing total capacity to 190,000 metric tons (209,300 short tons) per year. One solution would be for the municipality of Toulouse to incinerate the waste of smaller surrounding communities. To date, though the city has tried, this had not proved possible, partly for reasons of regional politics.

Table 2.3 shows the monthly performance of each furnace for the operating year 1974/75. According to these data, the average amount of waste incinerated in each furnace was 9.6 metric tons (10.6 short tons) per hour of use. This indicates that, when in use, the furnaces have operated well above their rated capacity of 8.15 metric tons (8.97 short tons) per hour. The explanation is that the furnaces are larger than necessary. The lower heating value on which size computations were based was 2,200 kcal/kg (3,970 Btu/lb). Municipal solid waste was analyzed in Toulouse during the acceptance tests for the first furnace in 1970. The average lower heating value was 1,750 kcal/kg (3,150 Btu/lb). This average appears to have remained stable until 1973, when the price of used paper climbed rapidly and a paper collection drive began (9). In

TABLE 2.3: OPERATION OF THE THERMAL COMPLEX
(first two furnaces)

Operating Period	Quantity of Waste Incinerated (metric tons*)	Incinerator Operation (hours)			Average Metric Tons/Hour	Turbo-Generator Operation (hours)
		First Furnace	Second Furnace	Total		
September 1974	7,756	711	0	711	10.9	713
October 1974	7,721	657	256	913	8.5	735
November 1974	7,052	253	512	765	9.2	700
December 1974	7,724	737	0	737	10.5	722
January 1975	7,791	417	517	934	8.3	720
February 1975	6,993	277	476	753	9.3	670
March 1975	7,191	89	658	747	9.6	668
April 1975	7,909	706	0	706	11.2	677
May 1975	7,864	730	67	797	9.9	725
June 1975	7,602	28	706	734	10.4	681
July 1975	6,806	0	730	730	9.3	654
August 1975	6,116	234	448	682	9.0	509
Total	**88,525**	**4,839**	**4,370**	**9,209**	**9.6**	**8,174**

SOURCE: Annual Report to the City of Toulouse by BETURE, 1974-75 *(Rapport Technique et Economique).*

* To convert metric tons to short tons, multiply by 1.1.

1974, the average lower heating value was down to 1,500 kcal/kg (2,700 Btu/lb). The average climbed up somewhat the following year, and during the acceptance tests for the third furnace in 1975, the average was measured at 1,560 kcal/kg (2,808 Btu/lb). In the last quarter of 1975, samples of waste from four different sections of Toulouse were analyzed. In the commercial center, the lower heating value was 1,570 kcal/kg (2,826 Btu/lb); in a central residential area, it was 1,650 kcal/kg (2,970 Btu/lb); in a single-family home area, it was 1,550 kcal/kg (2,790 Btu/lb); and in an area of new high-rise apartments, it was 1,570 kcal/kg (2,826 Btu/lb) (10). These values appear to have remained stable.

Since the actual lower heating value is smaller than the theoretical lower heating value, it is possible to surpass the capacity rating given by the manufacturer. (The actual lower heating value is low even for France. This is probably due to the provincial character of Toulouse.) In fact, the only real limitation is the amount of steam produced, which cannot exceed 23.4 metric tons (25.7 short tons) per hour (11). It has proved possible to incinerate waste at a rate as high as 11.2 metric tons (12.3 short tons) per hour, for example in April 1975 (see Table 2.3).

Table 2.3 shows also that the amount of waste treated each month never exceeded 7,900 metric tons (8,700 short tons) in the operating year 1974/75. This is less than the 10,000 metric tons (11,000 short tons) per month which represents the minimum necessary for continuous use of two furnaces, based on a minimum capacity of 7 metric tons (7.7 short tons) per hour for each furnace (12).

The potential heat which can be recovered is estimated at 1.01 kth/metric ton (3.65×10^6 Btu/short ton or 1823 Btu/lb) (13). This number is calculated from the average lower heating value of the waste and the thermal efficiency of the system. The average lower heating value has been estimated at 1,600 kcal/kg (2885 Btu/lb) and the thermal efficiency at 65% (14). The figure of 65% is based on the estimation that 25% of the theoretical potential is lost in the furnaces and in the production of steam, and a further 10% in the heat exchangers.

The amount of heat effectively recovered can be calculated by adding the electricity produced and the energy effectively used for heating and hot water. The electricity has to be translated into its thermal equivalent on the basis of 1 kWh = 1.1 therm to take into account the efficiency of the turbogenerator (15).

The amount of heat which could theoretically be recovered from the waste incinerated and the amount of heat actually recovered have been compared on a monthly basis for the operating year 1974/75 (see Table 2.4). To analyze the results one must distinguish between two periods:

— The heating period, from October to May, during which the amount of heat actually recovered depends

TABLE 2.4: ENERGY RECOVERY

Period of Operation	Quantity of Waste Incinerated (metric tons*)	Recoverable Heat (kth**)	Energy Recovered				Percentage Recovered	Heat Sold at Substation Level (kth)
			Electricity (kWh)	Hot Water (kth)	Central Heating (kth)	Total (kth)		
September 1974	7,756	7,865	305,000	1,084	2,261	3,680	46.8	1,453
October 1974	7,721	7,829	512,800	411	5,902	6,877	87.8	7,860
November 1974	7,052	7,150	496,300	0	4,188	4,734	66.2	12,981
December 1974	7,724	7,832	429,200	92	3,262	3,826	48.8	14,414
January 1975	7,791	7,900	583,400	532	4,885	6,059	76.7	16,200
February 1975	6,993	7,090	547,400	655	4,679	5,936	83.7	14,614
March 1975	7,491	7,595	451,000	860	3,994	5,350	70.4	13,143
April 1975	7,909	8,019	516,600	951	4,382	5,901	73.6	12,816
May 1975	7,864	7,974	340,600	567	3,690	4,632	58.1	5,088
June 1975	7,602	7,708	379,900	868	2,715	4 001	51.9	1,909
July 1975	6,806	6,901	339,600	593	2,488	3,455	50.1	1,378
August 1975	6,116	6,201	236,600	402	1,379	2,041	32.9	951
Total	**88,825**	**90,064**	**5,138,400**	**7,015**	**43,825**	**56,492**	**62.7**	**102,807**

SOURCE: Annual Report to the City of Toulouse by BETURE, 1974-75 *(Rapport Technique et Economique).*

* To convert metric tons to short tons, multiply by 1.1

** One kth (kilotherm) equals 3.97×10^6 Btu.

mostly upon the potential heat which could be recovered. The need for heating during this period is greater than the energy available from the waste.
— The nonheating period during which the amount of heat actually recovered depends mostly upon the need for hot water.

During the heating period in 1974/75, the potential heat which could be recovered from waste has been estimated at 61,400 kth (2.44×10^{11} Btu) and the heat actually recovered at 43,400 kth (1.72×10^{11} Btu) which gives a recovery rate of 71% (see Table 2.4). Some energy was lost because of insufficient demand for heating at the beginning and end of the period; consequently the heat was simply dissipated into the atmosphere. Some heat was also lost by keeping the condensers under pressure at a temperature of 100°C (212°F).

During the non-heating period of 1974/75, the potential heat which could have been recovered from waste has been estimated at 28,700 kth (1.14×10^{11} Btu) and the heat actually recovered at 13,200 kth (5.24×10^{10} Btu) (see Table 2.4), which gives a recovery rate of 46%. Energy was lost because the amount needed for hot water was less than the amount available. Thus, there is a market inefficiency as well as a thermal inefficiency.

In October 1975, shutters were added to the condensers which considerably increased the percentage of potential energy actually recovered. It has been estimated that 96% of the potentially usable energy was recovered in the month of February 1976 (16). To reach such levels in nonheating months would require an increased demand for hot water because of the addition of dwelling units, or the use of energy for air conditioning or industrial activities. However, at best only 65% of the energy content of the waste would be recovered, because of the limit of the thermal efficiency of the system.

System Reliability

The furnaces were designed to operate 24 hours a day, 330 days a year with a one-month period for routine maintenance. In reality, because of a shortage of waste to be treated during the operating year 1974/75, the two furnaces were in operation for a combined total of about 9,200 hours (see Table 2.3), which gives a level of activity of 58%.

Nonroutine maintenance costs amounted to approximately 1,300,000F ($261,000; 1969–1975) over a five year period, or 5% of the construction cost. These costs include all replacement parts. Maintenance costs were very low during the first three years and increased moderately during the next two years. It is, however, difficult to guess what would have happened if the two furnaces had operated at full capacity.

A landfill site is still maintained in the nearby town of Blagnac in case the incinerator should have to shut down. Thus far, however, this has never happened. In fact, the design of the system considerably reduces the risk of failure in the incineration of waste and the production of heat. The three furnaces are completely independent except for the overhead crane which introduces the waste, the ash pit for the incineration residues and the smokestack. The two oil-fired boilers of the heating facility can compensate for any decrease in heat recovery. The internal electrical system is coupled with the outside network which cuts in automatically in case of a breakdown.

A number of problems have been noted during the first five years operation:

- Since there is no preliminary crushing or compacting of the waste, large objects such as car engines or refrigerators are occasionally introduced, with the risk of damaging the furnace. Although the crane operator is supposed to watch the waste as it is unloaded from the trucks, bulky objects have slipped in a number of times and the furnace had to be stopped for hours.
- Some materials, especially aluminum, melt and stick to the rollers.
- Unless the condensers operate continuously (though not necessarily at full capacity) they corrode rapidly. One condenser has had to be replaced completely.
- Originally, a metal conveyor belt was used to carry the residues of incineration. This belt corroded, however, and a rubber belt was substituted.

Except for the change in design of the third furnace relative to the first two, most of the other changes in the system have been made to improve working conditions in the plant. The thermal complex now appears to be a modern, pleasant workplace.

Pollution Control

The volume of fluegas produced is 46,500 m^3 (1,642,000 cu ft) per hour per furnace (17). The gas passes through electrostatic precipitators, the efficiency of which exceeds 97% (18). It then escapes into the air through a 45 meter (148 ft) high smokestack.

The height of the smokestack had to be limited to meet the requirements of aviation safety because of the proximity of an airport. It contains two flues, the diameters of which are 4.4 and 2.4 m (14.4 and 7.9 ft). The first leads from the heating system and the second from the incinerator (19).

Current French legislation requires that particulate emission remain below 150 mg/m^3. The maximum permissible rate of CO_2 emissions is 7%; the maximum permissible rate of CO emissions is 0.1%.

Official pollution tests, performed during two separate months, gave the following results (20):

April 1970 (start of operations for the first two furnaces)

CO_2 = 12.7%

Particulates = 114 mg/m^3

CO = nil

October 1975 (start of operations for the third furnace)

CO_2 = 7.7%

Particulates = 75.5 mg/m^3

CO = nil

Additional tests were performed for the city of Toulouse by INSERM, the French Institute for Medical Statistics and Research *(Institut de la statistique et de la récherche medicale)* (see Tables 2.5 and 2.6). Their measurements were carried out near the top of the stack at a level of 40 m (131 ft) high for the gaseous emissions and 15 m (49 ft) high for the particulates. The results showed that all the particulates and gaseous compounds were in compliance except for HCl. The values found for the particulates ranged from 50 to 200 mg/m^3 with an average representative value of 100 mg/m^3. For the HCl, values ranged from 400 to 1,200 ppm with an average value of 750 ppm, which represents an emission of 5 g of HCl per kg of waste (0.005 lb HCl/lb waste) (21).

TABLE 2.5: ANALYSIS OF GASEOUS EMISSIONS

Component	Number of Tests	Concentration (ppm) Minimum	Maximum	Average	Observations
Cl_2	8	0	0	0	
$COCl_2$	10	<0.25	<0.25	<0.25	
Aldehydes & ketones	10	0.3	5	3	expressed as HCHO
H_2S	12	0	6	0	
Phenols	16	0	0.10	0.05	
CO	25	0	1,500	<100	
CO_2	25	50,000	120,000	85,000	
NO	20	45	120	75	
NO_2	10	0.2	1	0.4	
SO_2	20	50	120	100	
HF	20	4	15	8	
HCN	20	0	50	5	
Organic acids	20	40	250	100	expressed as CH_3COOH
HCl	20	400	1,200	750	
$PO_4^{\equiv}$	20	1	25	8	$PO_4^{\equiv}$ mg/m^3
NH_3	10	2	20	5	

SOURCE: Data gathered by INSERM, furnished by the municipality.

NOTE: Samples taken in the stack at a height of 40 meters (131 feet).

TABLE 2.6: PARTICULATE ANALYSIS

Component	Electrostatic Precipitator	Smokestack
Cl^-	4.45	22.00
$SO_4^=$	5.20	10.00
MO_3^-	0.001	0.07
F^-	0.15	0.20
$PO_4^{\equiv}$	2.40	1.40
SiO_2	37.0	1.80
C	8.3	7.50
Na^+	1.1	5.40
K^+	2.3	11.00
Ca^{++}	6.9	3.35
Mg^{++}	2.7	2.10
Pb^{++}	0.56	3.60
Fe	2.62	6.10
Zn^{++}	1.70	7.00
Ti^{++++}	0.55	0.40
Mn^{++}	2.20	0.11
Sn^{++}	0.15	0.30
Al^{+++}	8.92	0.50
Cu^{++}	0.072	0.22
Cd^{++}	0.030	0.10
Ag^+	0.006	0.01
Cr^{+++}	0.05	0.02

SOURCE: Data gathered by INSERM, furnished by the municipality.

NOTE: Sampling showed a concentration of particulates ranging between 50 and 200 mg/m^3 with an average concentration of 100 mg/m^3 Results are given in percentage of total particulates. Samples taken in the stack at a height of 15 m (49 ft).

The relatively high amount of HCl apparently results from the presence of PVC bottles, which make up 1% of the weight of refuse throughout France. (These bottles are used mainly for mineral water, vegetable oil and *vin ordinaire*.) At present, there is no regulation in France limiting the emission of HCl. However, the municipality seems to view this problem seriously and is trying to find a solution.

The problem of HCl emissions, and the fact that special exceptions were required for the limited height of the stack and for the use of residual fuel oil in the heating plant, have led the municipality (with the help of INSERM) to set up five monitoring stations at ground level. The analysis began in June 1976 and lasted for one year. The concentration of HCl in the atmosphere ranged from 3 mg/m^3 to 50 mg/m^3. Particulate Cl ranged from 2 mg/m^3 to 8 mg/m^3. HCl was chosen as a tracer for gaseous emissions and tin was chosen as a tracer for particles. However, the concentration of tin proved to be so low that it could not be traced with the instruments available. In any case, it was not possible to obtain a clear picture of the emissions resulting from the incinerator as background emissions were not recorded before construction began.

The combustion residues have been estimated at 30% by weight of the wastes incinerated, including 3.5% (by weight) of unburned material (22). These residues are landfilled (along with the untreated waste of a nearby town), but space is becoming a critical problem.

The results of the analysis at ground level will probably be presented to the regional environmental groups. The nearest residents complain quite frequently to the municipality, but it is difficult to say if these complaints are legitimate as there are never complaints about the old incinerator of Toulouse-Le Ramier which registers higher levels of emissions on tests.

Plant Building and Working Conditions

The thermal complex is situated on a 4.5 hectare (11 acre) tract in an industrial zone some ways from residential areas. The building itself is 121 m (397 ft) long and 57 m (187 ft) wide. Its general height is 17 m (56 ft), however, a large concrete block 36 m × 36 m × 30 m (118 ft × 118 ft × 98 ft) and a 45 m (148 ft) stack protrude (see Figure 2.3). The plant was designed by M. Alexis Josic, working under the supervision of M. Candilis, a well known architect, who designed the new town of Le Mirail where the plant is located.

FIGURE 2.3: PLAN OF THE THERMAL COMPLEX

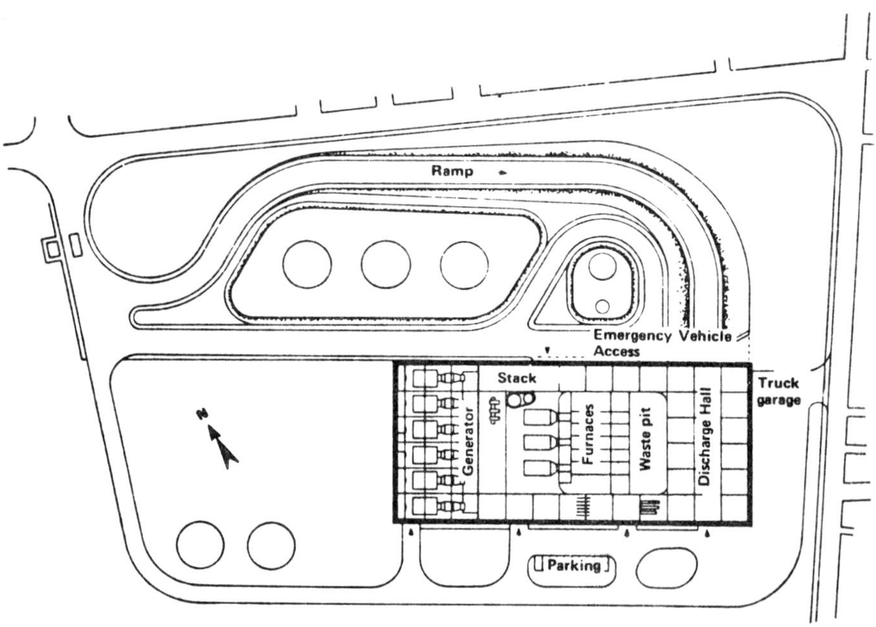

Several factors which are often neglected in industrial architecture have received careful attention: land use, organization of space, and grouping. The external aspect is somewhat heavy. The general impression is cold and colorless—even a bit crude—though pure and well balanced. Working conditions have been constantly improved since construction: overhead lighting in the central block, insulation for the furnaces, air conditioning, and a large window in the furnace control room have all been added.

Working conditions now appear satisfactory from the standpoints of ambient temperature, air quality, and building cleanliness. In certain places there is a problem with the noise level, but these are not permanent work posts. Deodorants are used to reduce the smell in the waste pit. The draft in the pit is still inadequate, probably because of the underutilization of the furnaces. The door leading to the tipping platform is now remotely controlled and is closed after the entry of each truck. This has reduced the emission of odors to the outside somewhat.

Only one major accident has occurred. It has led to the addition of an emergency stop on the conveyor belt which carries the incinerator residues.

Traffic is something of a problem since all the trucks tend to arrive during the same period (between 10:00 and 11:30 AM). This problem cannot be overcome for labor reasons, since it would require that the truck drivers change their breakfast time!

ECONOMICS AND MANAGEMENT

Management

The operation of the incinerator and heating facility, has been conceded to a private company SETMI, according to a system called *affermage*. The city of Toulouse retains ownership of the thermal complex which is operated by SETMI for a period of 20 years. SETMI was created specifically to operate this complex, the shareholders being the various companies which participated in the construction, including Tunzini (23). Other companies participating are SITECO which is in charge of technical control, and Est et Nord which has financial control.

The head of SETMI is a former technical adviser to the city of Toulouse. He is assisted by a director and a technical vice-director. Total employment is 47 people, most of them working throughout the thermal complex. However, five mechanics and one maintenance supervisor work specifically on the heating facility and the distribution networks (see Figure 2.4).

Among the 41 employees at the central unit, there are 5 executives, 13 maintenance workers, 20 workmen and 3 other administrative personnel. The workmen are organized in five shifts of 4 people working under the 3 × 8 sys-

FIGURE 2.4: ORGANIZATION CHART OF SETMI
(Operator of the Thermal Complex)

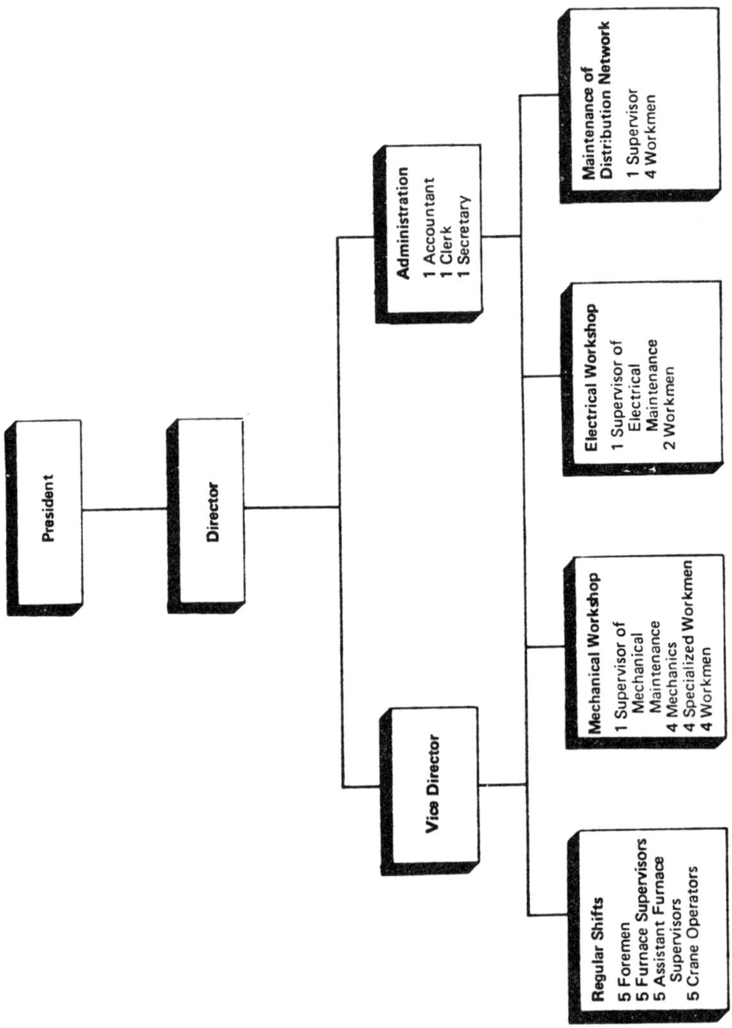

SOURCE: Derived by RPA from data provided by SETMI: M. Lartigue.

tem (3 teams working 8 hours each). There are fewer employees at Le Mirail than at other thermal plants of similar size in France.

The operation of the central unit is performed by 20 workers who incinerate between 88,000 and 100,000 metric tons (97,000 and 110,000 short tons) of waste annually. It would be possible to increase the amount of waste incinerated considerably [at least to 130,000 metric tons (143,000 short tons)] without increasing the personnel (24). In fact, SETMI is considering the transfer of incineration control from separate control centers for each furnace to the central control room in order to reduce the number of employees still further (25).

Certain jobs appear to demand considerable skill, particularly those of the furnace supervisor, who has to monitor the operation constantly, and the crane operator. It takes at least a year of experience to learn to supervise a furnace. The foremen have more than ten years experience in heating operations and a vice-director has 20 years experience in thermal operations. A foreman earns between 40,000F and 50,000F ($8,363-$10,460; 1976) per year, an unskilled workman earns 20,000F ($4,184; 1976) per year (26). A foreman's salary is average for the job category. The unskilled workmen earn the minimum wage.

The city exercises its overall control through the engineering firm BETURE which receives monthly reports from SETMI and presents an annual report to the city. This report includes technical and economic data but does not contain any financial information. Since SETMI is a private company, the financial statement, with the real costs of operations and their distribution, is not available to the city.

Capital Costs

The construction of the first two incinerators of phase I cost 25,200,000F in 1969 ($4,846,000; 1969 or $9,121,000; 1976) (see Table 2.7). Phase II, construction of the third incinerator, cost $17,700,000F in 1975 ($4,126,000; 1975 or $4,476,000; 1976) (27). A plant with equivalent capacity might cost less today, due to technical improvements leading to increased productivity.

TABLE 2.7: CONSTRUCTION COSTS

Phase I (1969)		**25,200,000F (59%)**	**$ 9,121,000–1976**
Two incinerators	16,800,000F (39%)		
Heat recovery equipment	8,400,000F (20%)		
Phase II (1975)		**17,700,000F (41%)**	**$ 4,476,000–1976**
Third incinerator	11,800,000F (27.5%)		
Heat recovery equipment	5,900,000F (13.5%)		
Total		**42,900,000F**	**$13,597,000–1976**

For phase I, out of a total cost of 25,200,000F, 16,800,000F was for the incinerators themselves and 8,400,000F for the heat recovery system (see Table 2.8). The contract was awarded not by the municipality itself, but by SETOMIP, the agency in charge of the new town development. SETOMIP received a subsidy from the national government of 3,500,000F (21%) for the incinerator (28).

TABLE 2.8: DISTRIBUTION OF CONSTRUCTION COSTS

Phase I (two incinerators–1969)	
Incinerators	
Construction	5,920,000F
Equipment	9,250,000F
Engineering services	960,000F
General expenses	560,000F
Total	**16,800,000F**
Heat Recovery System Equipment	
Construction	2,275,000F
Equipment	4,735,000F
Engineering services	430,000F
Debt service	700,000F
General expenses	300,000F
Total	**8,400,000F**

Phase II (third incinerator–1975)	
Incinerator and Heat Recovery Equipment	
Construction	885,000F
Equipment	14,103,000F
Engineering services	443,000F
Debt service	145,000F
General expenses	2,124,000F
Total	**17,700,000F**

SOURCE: M. Seman, SETOMIP

For phase II, out of a total cost of 17,700,000F, 11,800,000F was for the third incinerator and 5,900,000F for the heat recovery system (see Figure 2.4). This contract was awarded directly by the municipality which obtained a subsidy from the national government of 1,600,000F (14%) for the incinerator.

The subsidies from the national government (which amounted to 12% of the total cost of the thermal complex) were only for the incinerators and not for the heat recovery system. This may seem curious in a country where energy savings are considered highly important by the government. The project was funded essentially through the *Caisse des Dépôts et Consignations* (which plays the role of state banker for municipal projects) at the relatively low interest rates of 6.5% for phase I and 10.2% for phase II (see Table 2.9) (29).

TABLE 2.9: FINANCING

Phase I (two incinerators–1969)*	
3,500,000F	Subsidy for incinerator from national government
7,700,000F	Loan from SAMTAS at 8.75% for 15 years
8,740,000F	Loan from *Caisse des Dépôts et Consignations* at 6.5% for 20 years
7,960,000F	Loan from *Caisse des Dépôts et Consignations* at 5.25% for 20 years
Phase II (third incinerator–1975)**	
1,660,000F	Subsidy for incinerator from national government
16,040,000F	Loans from *Caisse des Dépôts et Consignations* at 10.2%

* SOURCE: M. Seman, SETOMIP
**SOURCE: M. Derbogossian, City of Toulouse

The amortization cost, including debt service (excluding subsidy), has been estimated by the municipality at 35F/metric ton ($7.40/short ton; 1975) incinerated in 1975 on the basis of 130,000 metric tons (143,000 short tons) incinerated (30). RPA estimates that for the first two furnaces only, the amortization cost would be 22F/metric ton ($3.80/short ton; 1969) incinerated.

Operating Costs

SETMI obtains three separate forms of revenue from the city of Toulouse.

- a fixed fee corresponding to fixed costs,
- a proportional fee corresponding to variable costs and proportional to the amount of waste treated,
- a fixed depreciation fee for major repairs and replacement.

These revenues are computed from an indexed formula which was set out in the initial contract. The fixed fee goes mainly for salaries and amounted to more than 1,200,000F ($264,000; 1974–1975) during the operating year 1974/1975 (see Table 2.10). The proportional fee corresponds to costs of materials and maintenance. It amounted to 1,100,000F ($242,000; 1974–1975) in 1974/75 or 12.4F/metric ton ($2.50/short ton; 1974–1975) incinerated. Finally the depreciation fee amounted to 680,000F ($149,500; 1974–1975) in 1974/1975. This fund is put into a special account and drawn on only when major repairs and replacements are actually needed. Credit is carried forward to the following year.

On September 1, 1975, the balance of this account was 1,400,000F ($326,000; 1975) (see Table 2.11) which indicates that to that date, no unexpected expense for major repairs had occurred. Total nonroutine maintenance costs for the operating year 1974–1975 were 571,243F (125,500), as shown in Table 2.12.

TABLE 2.10: DIRECT OPERATING COSTS
(francs, before tax)

Operation	1971/72	1972/73	1973/74	1974/75
Quantity of waste (metric tons*)	92,759.00	100,101.00	86,329.00	88,826.25
Heat recovered (kth**)	47,515.20	54,368.00	47,058.00	50,840.19
Heat recovered (kth/metric ton)	0.512	0.543	0.545	0.572
Gross cost of incineration				
–Fixed contribution	766,341.30	976,237.31	1,036,703.28	1,244,835.11
–Proportional contribution	730,490.23	872,666.69	866,484.62	1,098,811.48
–Contribution for major repairs	427,879.11	510,876.09	547,040.94	681,271.55
Total	**1,924.710.64**	**2,359,780.09**	**2,450,228.84**	**3,024,918.14**
Francs/metric ton	20.75	23.57	28.38	34.05
Price of heat recovered	806,993.58	984,194.00	1,425,853.46	2,361,214.06
Francs/kth	16.98	18.10	30.30	46.44
Net cost of incineration	1,117,716.86	1,375,586.09	1,024,375.38	663,704.08
Francs/metric ton	12.05	13.74	11.87	7.47

SOURCE: Annual Report to the City of Toulouse by BETURE, 1974-75 *(Rapport Technique et Economique).*

* To convert metric tons to short tons, multiply by 1.1.
** One kth (kilotherm) equals 3.97×10^6 Btu.

Conversion Rates:

	F/$
1971/72	5.27
1972/73	4.75
1973/74	4.63
1974/75	4.55

TABLE 2.11: NONROUTINE MAINTENANCE ACCOUNT (francs)

Operating Year	Receipts from City of Toulouse	Expenditures	Conversion Rates (F/$)
1969/70	189,318.96		5.36
1970/71	385,149.85	137,840.71	5.52
1971/72	427,879.11	161,448.00	5.27
1972/73	510,876.09	92,213.01	4.75
1973/74	547,040.94	331,708.97	4.63
1974/75	681,271.55	571,242.85	4.55
Total	**2,741,536.50**	**1,294,453.54**	
Balance as of Sept. 1, 1975	**1,447,082.96**		

SOURCE: Annual Report to the City of Toulouse by BETURE, 1974-75 *(Rapport Technique et Economique).*

TABLE 2.12: NONROUTINE MAINTENANCE COSTS FOR OPERATING YEAR 1974–1975 (francs)

Furnaces		
–Mechanical parts	48,730.97	
–Electricity	7,815.95	
–Repair of flues	40,320.90	
Waste crane	18,718.45	
Ash crane	5,749.86	
Ash conveyor belt	220.42	
Electrostatic precipitators	84,346.45	
Rebuilding of flues and installation of ground-level ventilators*	200,000.00	
Turbo-generator	20,080.15	
Compressors	1,239.32	
Air condensers	3,418.80	
General electricity	4,518.55	
Skilled labor	82,094.65	
Insurance premium	24,988.38	
Total	**571,242.85**	**($125,500)**

SOURCE: Annual Report to the City of Toulouse by BETURE, 1974-75 *(Rapport Technique et Economique).*

* Work carried out in July, 1974, but carried on accounts of Operating Year 1974-75

The total amount paid by the City of Toulouse to SETMI was approximately 3,000,000F ($659,000; 1974–1975) during the operating year 1974/1975. (See Table 2.10). In return, the city receives from SETMI a sum proportional to the amount of heat which has been recovered. This sum is also computed from a formula set in the initial contract and based on the price of No. 2 oil. This repayment amounted to 2,400,000F ($527,000; 1974–1975) for 1974/1975 or 46F/kth ($2.54/$10^6$ Btu; 1974–1975) (see Table 2.10). A market for the residues of incineration has not yet been found.

Thus, while the total amount paid by the municipality to SETMI in the three categories was appoximately 3,000,000F during 1974/1975 [or 34F/metric ton ($6.80/short ton; 1974–1975) incinerated] the total amount received by the municipality from SETMI during the same period was approximately 2,400,000F ($527,000; 1974–1975) [or 27F/metric ton ($5.40/short ton; 1974–1975) incinerated]. On balance, the costs incurred by the municipality of Toulouse for the operation of its incinerator during 1974/1975 can be estimated at 7.5F/metric ton ($1.50/short ton; 1974–1975). This figure has been computed without tax considerations (17.6%), and is equivalent to approximately 9F/metric ton ($1.80/short ton; 1974–1975) including taxes (31). Based on equivalent computations, RPA has developed the following estimates for overall direct operating costs (see Table 2.10).

12F/metric ton ($1.96/short ton) in 1971–1972
13.7F/metric ton ($2.61/short ton) in 1972–1973
11.9F/metric ton ($2.33/short ton) in1973–1974
7.5F/metric ton ($1.49/short ton) in 1974–1975

The reduction in cost is due mainly to the increase in the value of the heat recovered as a result of the general increase in the price of energy.

Overall Costs

Considering the first two incinerators only, the total cost of incineration to the municipality can be obtained by adding the cost of amortization, or 34F/metric ton ($6.80/short ton; 1974–1975) to the direct operating cost, or 7F/metric ton ($1.40/short ton; 1974–1975). This gives a total of 41F/metric ton ($8.20/short ton; 1974–1975) incinerated in 1975. However, this figure should be considered as only a rough estimate for at least four reasons:

- In addition to the revenues received from the city (see Table 2.10) SETMI (the operator of the thermal complex) receives revenue for which the city receives no credit, from the sale of heat and hot water to the households of Le Mirail. In the operating year 1974/1975, SETMI charged 72.7F/kth ($4/$10^6$ Btu) before taxes [90F/kth ($5/$10^6$ Btu) including taxes]. This price was at or below the market price of such substitutes as fuel oil and gas, and may thus represent a small subsidy to the homeowner.
- The cost of the distribution network is not included.

- A subsidy of about 12% of total cost was obtained from the national government, moreover interest rates are relatively low.
- No cost for the land on which the incinerator was built has been taken into account as it belonged to the municipality.

However, these estimates give a first approximation of the total costs of incineration at Toulouse.

CONCLUSIONS

A new town is probably the optimal location for a waste-to-energy system. The decision to recover energy at Toulouse-Le Mirail in the form of heat rather than electricity (except for in-plant use) seems to have been wise. There is a state-owned monopoly for power generation in France (*Electricité de France*) which provides electricity at relatively low rates. Furthermore, the recovery of heat is more efficient than the recovery of electricity would be.

This system has no apparent fundamental technical problem. The VKW furnaces are simple and robust, although they have never been tested to full design capacity. Incineration, recovery of energy, and provision of domestic heat and hot water are well integrated.

The thermal complex operates relatively inexpensively. This is due, in part, to the decision not to generate electricity and also, in part, to the low cost of unskilled labor in France. Operation would be still less expensive if more waste were incinerated and if more markets for the recovered energy could be found. The relationship between the City of Toulouse and the private company which operates the thermal complex appears to be very good and cost-effective.

Construction of the third incinerator while the first two were operating well below capacity was a mistake. Provision had been made for such a facility in the Sixth National Plan, a 14% subsidy was available, and inertia prevailed.

REFERENCES

Notes

(1) Personal communication with M. Cazaux, economist, Departmental Division of Public Works, Toulouse.

(2) Noirot, "L'Usine du Mirail-Traitement des Ordures," *Ingénieurs des Villes de France,* May 1976.

(3) Personal communication, M. Bressand, Technical Services, City of Toulouse.

(4) Internal report of the City of Toulouse.
(5) Circular of October 18, 1965.
(6) Personal communication, M. Roger, Director of Technical Services, City of Toulouse.
(7) All technical data on the incinerator are taken from:

- Noirot, *op. cit.*
- Information provided by the contractor, the engineering firm of Tunzini
- Information gathered on a visit to the thermal complex by RPA staff

(8) All technical data on the heating system are taken from:

- Noirot, *op. cit.*
- Information gathered on a visit to the thermal complex by RPA staff

(9) *Ministère de la Qualité de la Vie.*
(10) BETURE
(11) Personal communication, M. Noirot.
(12) BETURE, Annual Report to the City of Toulouse, 1974–1975. *("Rapport Technique et Economique").*
(13) Personal communication, M. Noirot.
(14) BETURE, *op. cit.*
(15) *Ibid.*
(16) BETURE, *op. cit.*
(17) INSERM
(18) Information furnished by contractor, confirmed by City of Toulouse.
(19) Information furnished by contractor.
(20) *Association des Producteurs d'Appareils à Vapeur et Electriques.*
(21) INSERM
(22) BETURE, *op. cit.*
(23) The shareholders are *Tunzini, Cie Générale de Chauffage à Distance, SITECO, Cie Générale des Eaux, Industrielles d'Exploitation de Chauffage, Streichenberger* and *Est et Nord.*
(24) Personal communication, M. Lartigue, SETMI.
(25) *Ibid.*
(26) *Ibid.*
(27) Personal communication, M. Derbogossian, City of Toulouse.
(28) Personal communication, M. Derbogossian.
(29) Personal communications, M. Derbogossian, and M. Seman, SETOMIP.
(30) Estimate developed by the City of Toulouse; personal communications, M. Roger, M. Bressand.
(31) Noirot, *op. cit.*

Persons Contacted

The following persons have cooperated with RPA in preparing this case study:

M. Noirot, Technical Engineer, City of Toulouse

M. Roger, General Director of Technical Services, City of Toulouse

M. Bressand, Chief Engineer, City of Toulouse

M. Lartigues, Assistant Director, SETMI

M. Derbogossian, Director of Finance, City of Toulouse

M. Vétu, Commercial Director, Tunzini, Paris

M. Mast, Technical Director, SITECO, Paris

M. Leprêtre, Operations Engineer, Tunzini, Paris

M. Cazaux, Economist, Departmental Division of Public Works, Toulouse

M. Séman, Financial Department, SETOMIP, Toulouse

M. Colardeau, Engineer, Ministry of the Environment, Paris

Case Study of Geneva-Cheneviers, Switzerland

Material for this chapter has been based upon a report prepared by Resource Planning Associates, Inc., Washington, D.C. and Paris, and Consultex, SA, Geneva (CONS/2103-2).

SUMMARY

The City of Geneva, population 159,000, is the administrative center of the Canton of Geneva, population 340,000. The Canton owns a number of facilities for the treatment of waste. The individual municipalities are responsible for waste collection.

Geneva's chief waste treatment facility is the Cheneviers incinerator. Two Von Roll integrated boiler incinerator furnaces have a rated capacity of 200 metric tons (220 short tons) per day each. Superheated steam at 360°C (680°F) and 32.4 bars (32 atm) powers a 6200 kW turbo-generator unit. The electricity is sold to the cantonal grid.

Total incinerable waste in the Canton of Geneva has varied from 120,000 to 130,000 metric tons (132,000 to 144,000 short tons) annually during the last five years. For the last two years, total and per capita tonnage have been declining. Per capita incinerable waste was 363 kilograms (800 lb) in the year 1975, of which 257 kilograms (606 lb) were household waste. Eighty-seven thousand, five hundred metric tons (96,386 short tons) of this waste was burned in the Cheneviers incinerator in 1975. The remainder was landfilled, due to the lack of capacity at the incinerators.

The system which began operating in 1966, cost approximately 40 million Swiss Francs ($9.3 million; 1965 or $23 million; 1976). Three-quarters of this sum was for land, buildings, construction and equipment. A large station and dock for the transfer of waste accounted for the remainder. Municipalities paid the Canton at the rate of 84 Swiss Francs per metric ton ($29.56 per short ton) of solid waste treated in 1975. The cost per metric ton of waste treated at the Cheneviers plant in 1974 was 64.25SF ($19.57 per short ton), including amortization.

The Von Roll design of this plant is now out of date. Extensive modifications were made to correct corrosion problems in the furnace.

BACKGROUND

Economic and Demographic Data

The City of Geneva is the capital of the Canton of Geneva, which is one of the major political subdivisions of Switzerland. The canton is situated in the extreme southwestern corner of Switzerland, nestled between the Jura mountains on the north and the Alps on the south. The canton is shaped roughly like a U wrapped around the southwestern tip of Lake Leman. The Rhône River begins at the tip of this lake and splits the canton into two parts.

The population of the City of Geneva as of December 31, 1975, was 158,698. The population of the rest of the canton (including 44 other municipalities) was 181,158, for a total for the whole canton of 339,856. Nine towns (with a population of 6,086) in the neighboring Canton of Vaud are also served by the solid waste plants in the Canton of Geneva.

The total area of the Canton of Geneva is 282 square kilometers (109 square miles) and the population density is 1,205 per square kilometer (3,118 per square mile).

The Canton of Geneva is one of the wealthiest areas of the world. According to the Union de Banques Suisses, the total income of the canton in 1974 was $2.856 billion, giving a per capita income of $8,378.

Geneva is a center for light industry: machinery manufacture, food processing, specialty chemicals production, and watchmaking. There is no heavy industry or major polluting industry in the canton. The canton's largest sector is the service sector, including banking, hotels, restaurants, and commercial retailing.

Waste Collection and Treatment

The amount of solid waste generated in the Canton of Geneva and the nine towns in the neighboring Canton of Vaud for the years 1971 to 1975 is shown in Table 3.1.

TABLE 3.1: SOLID WASTE GENERATION IN THE CANTON OF GENEVA (metric tons)*

Type of Waste	1971	1972	1973	1974	1975
Household waste	94,633	95,443	99,870	92,992	88,857
Incinerable industrial waste					
Ordinary waste	11,209	13,429	17,349	20,990	23,183
Voluminous waste	522	192	458	945	925
Confidential papers, etc.	247	302	224	238	215
Used hydrocarbons	579	545	615	708	737
Dried sewage sludge	11,030	10,395	9,478	10,253	8,046
Sewage screen waste	1,715	1,461	1,904	2,498	2,201
Dangerous waste	839	673	742	999	1,243
Total	26,141	26,997	30,770	36,631	36,550
Total incinerable waste	120,774	122,440	130,640	129,623	125,407
Non-incinerable industrial waste	17,568	20,348	17,860	23,096	19,960
Total solid waste	**138,342**	**142,788**	**148,500**	**152,719**	**145,367**
Household waste per capita (kg/yr)**	244	275	278	267	257
Incinerable waste per capita (kg/yr)	350	353	363	372	363
Total solid waste per capita (kg/yr)	401	411	413	439	420
Population (as of 12/31)	345,053	347,379	359,959	348,133	345,942

* To convert metric tons to short tons, multiply by 1.1.
**To convert kilograms to pounds, multiply by 2.2.

Household waste rose from nearly 95,000 metric tons (104,650 short tons) in 1971 to a peak of nearly 100,000 metric tons (110,000 short tons) in 1973, and then started a significant decline to reach only about 89,000 metric tons (98,000 short tons) in 1975. This decline coincided with a small decrease in the population of the Canton, but the per-capita generation of household waste reflected the overall drop, falling from a high of 278 kilograms (613 lb) of waste per inhabitant per year in 1973 to only 257 kilograms (567 lb) of waste in 1975. This decrease may have been caused by the economic recession.

Incinerable industrial waste is listed in detail for the same years. This category has shown rather more consistent growth, with only a slight decline in 1975. Total incinerable waste, therefore, reflects the preponderance of household waste which also peaked in 1973.

The variation in population was such that the quantity of total incinerable waste (household and industrial) per capita peaked one year later, in 1974, with a noticeable fall in 1975. Total and per-capita solid waste generation, including nonincinerable waste, also peaked in 1974 and fell in 1975.

Collection and Disposal Procedures

Swiss federal law leaves the organization of solid waste collection and treatment to the individual cantons to legislate and control. At the time of the major investment program in the Canton of Geneva that resulted in the building of the Cheneviers plant, the canton passed new solid waste legislation, leaving the collection of waste to the individual municipalities, but organizing and operating all solid waste treatment at the cantonal level.

Under cantonal law, each municipality is responsible for the collection of its own household waste, including all financing and actual provision of services. The municipalities are told which of the several treatment plants are their delivery destinations and when such deliveries are to be made. Industrial and commercial waste is also treated by the cantonal authorities but private generators are responsible for the delivery of their own waste to the treatment centers.

During the last 16 years, the canton has set up a number of solid waste treatment centers. There is a major incinerator at Cheneviers where electricity is generated. There is also a smaller incinerator without any energy recovery at Richelien, a compost plant at Villette, a very small incinerator for special waste at Nant D'Avril, and a sanitary landfill.

The methods of treating the solid waste generated in Geneva during the last 5 years is shown in Table 3.2. Of the incinerable waste from both household and industrial origin, from 64 to 79% has been incinerated or turned into compost; the remaining incinerable waste has been landfilled.

In addition, a significant amount of nonincinerable waste and a considerable volume of incinerator ash have also been landfilled. Thus, incineration has made a major contribution to the reduction of the total volume, but there is still a considerable amount of material that has to be landfilled, including an important volume of incinerable material.

TABLE 3.2: SOLID WASTE TREATMENT METHODS IN THE CANTON OF GENEVA (metric tons)*

Treatment Plant	1971	1972	1973	1974	1975
Cheneviers incinerator	70,568	67,763	82,007	80,093	87,507
Richelien incinerator	6,027	5,223	6,775	5,938	5,441
Villette compost plant	5,475	5,617	5,567	6,433	5,882
Nant D'Avril incinerator	–	–	742	644	739
Sanitary landfill	56,272	64,187	53,410	59,611	45,828
Total primary treatment	**138,342**	**142,790**	**148,501**	**152,719**	**145,397**
Incinerable waste incinerated (%)	68	64	73	72	79
Incinerable waste landfilled (%)	32	36	27	28	21
Waste directly landfilled	56,272	64,187	53,410	59,611	45,828
Incinerator ash landfilled	29,636	25,765	32,307	32,179	33,373
Total landfilled	**85,908**	**89,952**	**85,717**	**91,790**	**79,201**

*To convert metric tons to short tons, multiply by 1.1.

The canton law establishes the general principle that the generator of the waste must pay for its disposal. Industrial generators pay a fixed fee per ton, depending upon the category of waste. These fees are reviewed each year. The municipalities also pay for the treatment of the waste generated by their inhabitants, but from indirect taxation on income, rather than in direct proportion to the amount of waste generated by each individual.

The canton operates all of its treatment plants on a full cost recovery basis—hence, as a nonprofit enterprise. There are no subsidies, direct or indirect, in this approach. Each year the authorities forecast the likely total costs and set the rates that the industrial generators will have to pay during the year. At the end of the year, the income from the industrial generators and the other income from the compost and energy recovered is subtracted from the total aggregate costs of all of the treatment plants and the remainder is divided by the total amount of waste from the municipalities to establish a per ton charge for household waste. Each town is then charged according to the amount of waste generated and delivered for treatment.

One additional element in the treatment system is the transfer of waste from trucks to barges at a station situated near the center of the city. Most of the municipal waste is transferred here and shipped down the Rhône River to the Cheneviers plant where the incinerator is fed with waste directly from the barges. The cost of this transfer operation is included in the total aggregate charge paid by industrial and municipal clients.

The evolution of the charges for waste treatment since the Cheneviers plant began operation in 1966 are shown in Table 3.3. The charges indicated at the top of this table are the charges per ton for industrial waste treated by the canton. The row for household waste shows the difference between the gross costs and the net costs of all income before the municipalities are charged.

In addition, a special fee is added to the charges of the municipalities to round-off the fees and to bring them a bit closer to what industrial users have to pay. These fees go into a special fund for major repairs. Starting in 1974, a second special fund was established. Both industrial and municipal users contribute on a per ton basis to provide bridging interest payments for the new extension of the Cheneviers plant which is now being built.

Over the years, the charges for waste treatment have increased to all users. The charge to industries has increased from 45 Swiss Francs per metric ton ($9.43 per short ton) in 1966 to 80SF per metric ton ($28.15 per short ton) in 1975, with an additional 15SF ($5.81) special charge in 1975. (The devaluation of the U.S. dollar during this period accounts for the greater increase in the dollar figure.)

The amount paid by the municipalities for their waste treatment has gone from 43.23SF per metric ton ($11.15 per short ton) in 1966 to 84.00SF per metric ton ($29.46 per short ton) in 1975.

The cost of collection and transport of waste is not clearly known and varies from one town to another. It is estimated to be at least equal to that of the treatment charge, however, and this would mean that the total cost of collection and treatment of solid waste from households in the canton in 1975 is about 170SF per metric ton ($59.90 per short ton).

TABLE 3.3: EVOLUTION OF TREATMENT CHARGES AND COSTS FOR SOLID WASTE DISPOSAL SERVICES IN GENEVA (in Swiss Francs per ton)*

Charge/Cost	1966 (6 mo)	1966/67 (12 mo)	1967 (6 mo)	1968	1969	1970	1971	1972	1973	1974	1975
Industrial waste											
Ordinary burnable waste	–	45	–	50	–	–	–	–	60	70**	80**
Voluminous waste	–	70	–	70	–	–	–	–	80	80**	90**
Waste placed directly in the furnace	–	90	–	90	–	–	–	–	90	90**	100**
Used hydrocarbons	–	50	–	50	–	–	–	–	10	10	10
Nonputrifying waste	–	10	–	10	–	–	–	–	10	15	15
Household waste	18.71	53.23	56.25	48.96	46.95	46.762	57.59	62.89	52.21	55.17	64.08
Fund for repair parts	–	–	–	1.04	3.05	4.238	–	–	10.79	8.83	4.92
Special fund**	–	–	–	–	–	–	–	–	–	10.00	15.00
Total paid by municipalities	18.71	53.23	56.25	50.00	50.00	51.00	57.59	62.89	63.00	74.00	84.00

*1966-75 conversion rates:

Year	SF/$
66	4.33
67	4.33
68	4.30
69	4.31
70	4.31
71	4.11
72	3.82
73	3.17
74	2.98
75	2.58

**The special fund is to pay interest charges on the costs of the extension of the Cheneviers plant before this extension begins operating. The categories of industrial waste marked with ** are also charged a supplementary fee at the same rate as the household waste.

Political Structure of Solid Waste Control

The canton's waste treatment centers are administered and controlled by an executive body consisting of seven State Counselors, each of whom is in charge of one of the seven major departments (see Figure 3.1). One of the executive committee members is thus also head of the Department of Public Works, which is comprised of several Directorates—one of which is called Genie Civil (i.e., infrastructure—roads, bridges, and the like). This directorate has two major subgroups, one of which is the Division of Operations and Maintenance, which in turn has two operational services. One of these operational services is in charge of Assainissement (Sanitation). This service supervises both solid and liquid waste treatment plants in the canton, and one of these solid waste plants is the incinerator at Cheneviers.

DEVELOPMENT OF THE SYSTEM

The Swiss are reputed to be intellectually innovative and financially cautious. They have a tradition of participatory democracy which keeps a large amount of decision making at the local level.

Ever since the nineteenth century, there has been proposals by the City of Geneva to build an incinerator for eliminating waste. In 1899, the city voted funds for such a plant, but the idea was allowed to die. Various other proposals were put forward in the 1930s and again in the 1950s, but none of these were brought to completion.

The Canton of Geneva is a rather small area. In the last few decades it had already become apparent that there were relatively few areas in the canton available to dump untreated solid waste. Thus there was pressure to reduce the dumping of waste. Incineration was an obvious possible solution.

In 1960, the Cantonal Engineer, Mr. Yves Maystre, and the Public Works Department staff studied the problem of waste and sludge for the entire canton. Their report led to a special commission composed of members of the Department of Public Works, representatives of the City of Geneva, and representatives of other municipalities of the canton which also wanted to participate in an eventual project. This group proposed that money be allocated to study a mixed treatment plant. Such a facility would have provided composting and incineration, treatment of sewage sludge, and the recovery of electrical energy. The site was to be Verbois on the Rhône River and the waste was to have been transported by barge. In 1962, the Department of Public Works published two studies to explain and to solicit support for this project. The proposal was accepted and money was allocated by the City of Geneva, by several municipalities, and by the Grand Council of the canton.

The details are now fading from memory, but most commentators agree that the major force behind the acceptance of the project was the political

FIGURE 3.1: ORGANIZATION CHART OF THE CANTON OF GENEVA FOR SOLID WASTE CONTROL

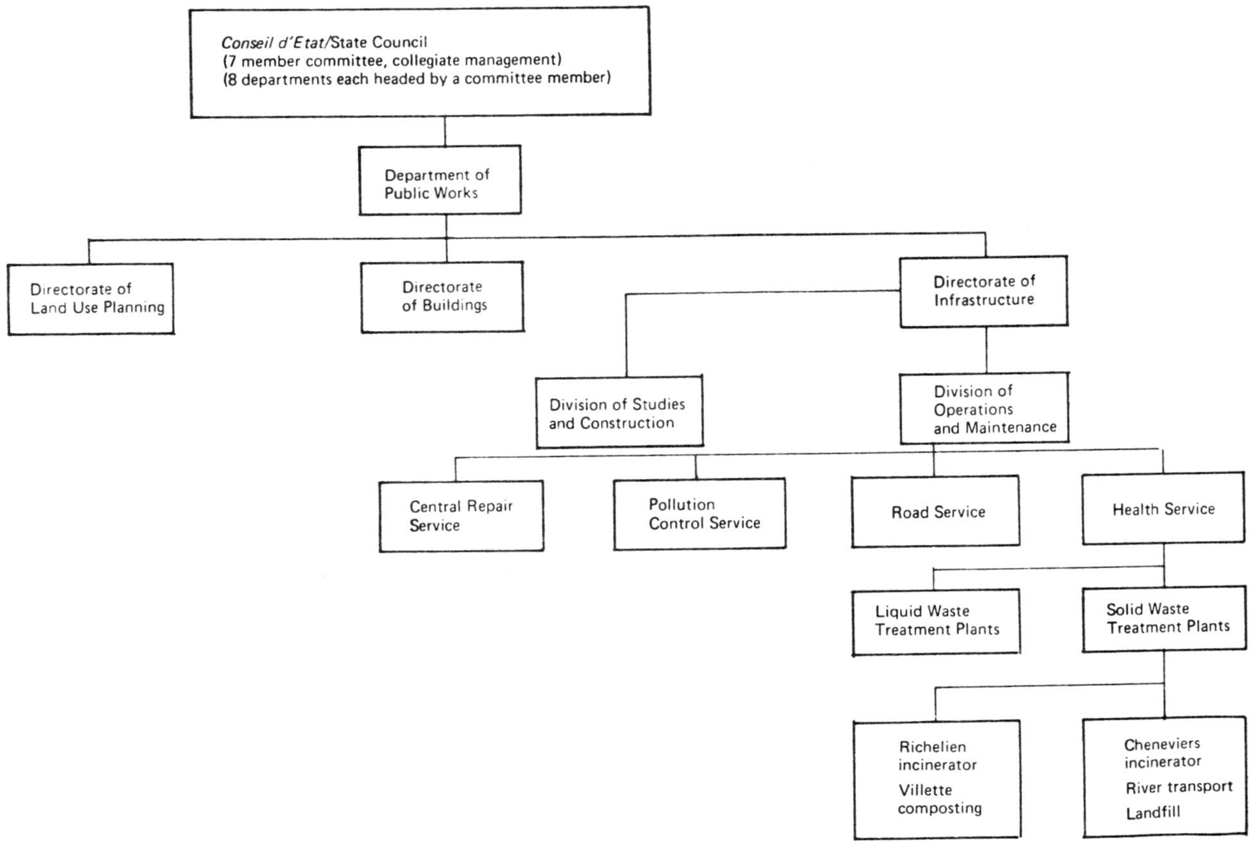

negotiating skill of Mr. Maystre. As mentioned above, the critical element in the planning process was the preparation of the solid waste analysis which Maystre and the cantonal Public Works Department produced in 1960. This sparked heated debate and led to the formation of the larger study commission which looked at specific options and finally recommended the incinerator and electricity recovery plant.

There was spirited debate about the type of system that should be chosen. Some felt incineration was the only sensible solution, while the agricultural lobby argued for a system that would produce soil conditioner.

About this time, as the canton had not yet decided to take over all solid waste treatment, several municipalities decided to build and operate a small compost plant. Another group of municipalities decided on a small incinerator without heat recovery.

The ultimate decision was to build a hybrid system. It was agreed that the Cheneviers plant would be built first with incineration and recovery of electricity, but with room for eventual composting plants if results from the small composting plant mentioned above looked promising.

This decision set arbitrary limits on the size of the plant, and resulted in a smaller than optimal plant being built. The final compromise called for the plant to incinerate ⅔ of the waste and to produce compost from the rest. However, the small composting plant produced a low quality compost and little could be sold. Most was hauled off to the dump. The composting plant at Cheneviers was never built.

Having decided on an incinerator and electricity recovery, the canton solicited bids. Von Roll (of Zurich) was selected to be the major supplier of equipment, in combination with Escher Wyss for the boiler and Sulzer for the generator. Architects and builders in Geneva received the construction contracts. It is impossible to reconstruct the decision to select Von Roll but the choice was defensible both technically and nationalistically, and these were the factors that predominated.

The Geneva authorities placed their confidence in Von Roll for the specific design of the plant. However, the plant that Von Roll proposed and the canton accepted involved a novel arrangement of the furnace and boiler that eventually resulted in major operating and maintenance problems. This will be discussed later in the report.

The canton in early 1960s was sufficiently affluent to permit a most ambitious project. Innovative architecture was approved, at rather substantial cost. There were hardly any legal environmental standards at the time, but some were self-imposed. The canton had a financial surplus and provided all of the financing at very favorable rates of interest and with a long amortization period.

The option to have an incinerator without any energy recovery did not receive much attention. The reasons for this omission seem to be bound up with political, social, and emotional imperatives. The canton has recently decided to expand its incinerator and again will add an electricity generation unit. It is by no means clear that this capacity makes sense economically or financially, but other considerations seem to make energy recovery essential.

TECHNICAL INFORMATION

The Basic System

The incineration plant at Cheneviers in Geneva consists of two Von Roll waterwall furnaces side by side, fed by a single overhead crane. Both furnaces power the same turbo-generator set and share a common stack. Figure 3.2 and Table 3.4 is a simplified schematic outline of the major components of the plant and one furnace. Details of the source and characteristics of the major equipment are also shown therein.

The waste arrives either by truck or by barge. There is a small bin (covered to prevent odor escaping) for waste received by truck. Barges that arrive from the transfer station 10 miles away serve the dual purpose of transport and storage. The barges are hauled right into the plant, next to the truck deposit pit, and serve as an alternative source of supply of furnace input. Oversized items are picked up by the crane and dropped into a third storage bin where there is a Von Roll shredder.

The crane is operated from the control room. The operator can see into each loading chute with a television system and thus check the requirements of each furnace.

The waste falls into a hopper and is moved by a vibrating table onto the drying grate. Movement in the drying grate gradually advances the waste to the point where it falls onto the main grate for burning. This grate has a combination of fixed and moveable elements to move the waste forward. There are also three moveable rakes which rise and fall to agitate the waste for more complete incineration. Burn out takes place at the end of the combustion grate. All the grates are constructed of refractory iron.

After reaching the end of the burning grate, the remaining slag falls into a quenching pit from which it is continuously extracted by an endless chain conveyor and moved to an ash pit. Eventually it is lifted from this pit to a truck and hauled to the controlled landfill.

In the original design of the plant, air was injected beneath the drying and burning grates. (See later discussion of system modifications.) The air then rose through the main combustion chamber, past lateral and frontal waterwalls, down through a second passage filled with more steam tubes, passing the

FIGURE 3.2: SCHEMATIC DIAGRAM OF THE CHENEVIERS (GENEVA) INCINERATOR

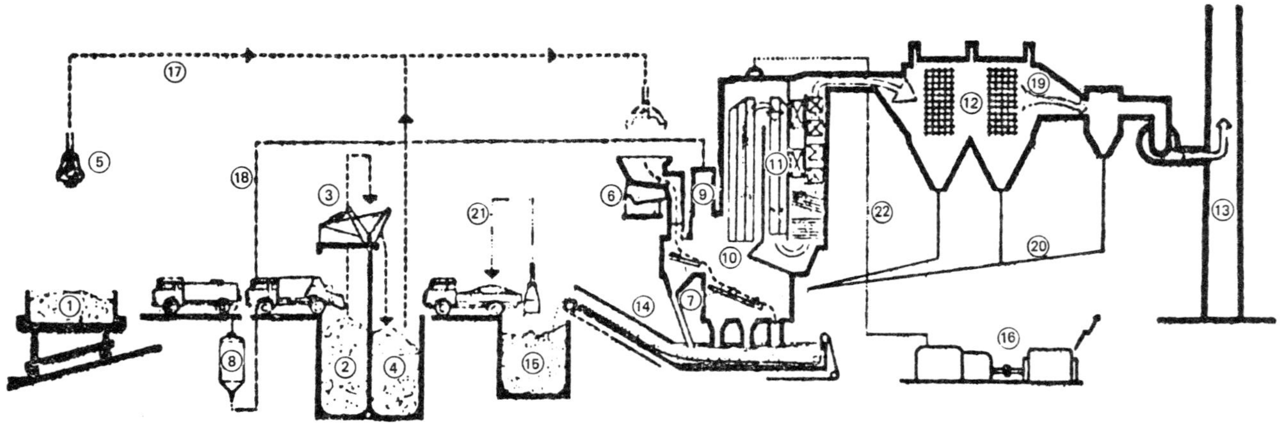

1. Barge
2. Storage pit for oversized waste
3. Shredder for oversized waste
4. Storage bin for household waste
5. Loading crane
6. Vibrating hopper
7. Drying and burning grates
8. Waste oil decantation
9. Waste oil burners*
10. Combustion chamber
11. Boilers
12. Electrostatic precipitators
13. Smokestack
14. Ash evacuation
15. Ash storage bin
16. Turbo-generator
17. Path traveled by waste
18. Path traveled by waste oil*
19. Path traveled by combustion gases
20. Recovery of ash from gases
21. Ash
22. Steam flow

*Discontinued.

superheaters en route, and finally up through a third passage where a split channel led it either past the economizer or the air preheater. The gases then were drawn into the main electrostatic precipitator bank and finally through a cyclone unit before reaching the induced draft fan and going up the stack.

TABLE 3.4: CHENEVIERS (GENEVA) INCINERATOR

Technical Data

Component	Manufacturer	Characteristic	Metric System	English Equivalent
Barge		Length	43 m	141 ft
		Width	8.63 m	28 ft
		Deadweight	125 metric tons	137.5 short tons
		Capacity (mass)	120 metric tons	132 short tons
		Capacity (volume)	600 m^3	785 cu yds
Shredder for oversized waste	Von Roll	Throughput	200 m^3/hr	262 cu yds/hr
		Shear force	72 metric tons	79.2 short tons
		Opening	3.8 x 4 m	12.5 x 13.1 ft
Storage bin for household waste		Capacity	2,600 m^3	3,400 cu yds
Loading crane	Von Roll	Mass	7 metric tons	7.7 short tons
		Volume	3 m^3	4 cu yds
Vibrating hopper	Schenk	Length	3.5 m	11.5 ft
		Width	2.5 m	8.2 ft
		Capacity	15 metric tons/hr	16.5 short tons/hr
Drying and burning grates (refractory iron area)	Von Roll	Drying	3.01 x 3.3 m	9.8 x 10.8 ft
		Burning	3.01 x 5 m	9.8 x 16.4 ft
Waste oil decantation			4 reservoirs	
Waste oil burners*		Capacity	250 kg/hr	551 lb/hr
Boilers	Escher-Wyss (Eckrohr)	Heating surface/boiler	835 m^2	8,984 sq ft
		/superheater	193 m^2	2,077 sq ft
		/economizer	340 m^2	3,658 sq ft
		/air preheater	1,115 m^2	11,997 sq ft
		/superheater exit pressure	32.4 bars	32 atm
		/temperature	375° C	707° F
		Steam production	17 metric tons/hr	18.7 short tons/hr
		Maximum steam production	24 metric tons/hr	26.4 short tons/hr
		Refractory brick	160 metric tons	176 short tons
		Insulation brick	32 metric tons	35 short tons
Electrostatic precipitators	Svenska Flaktfabriken	Maximum particle content	300 mg/m^3	
		Power	60,000 V, 300 milliamps	
Smokestack		Height	103 m	338 ft
		Width, base	4.88 m	16 ft
		Width, top	3.1 m	10.1 ft
Ash evacuation		Length of endless chain	60 m	197 ft
Ash storage bin		Volume	800 m^3	1,046 cu yds
Turbo-generator	Escher-Wyss turbine (mono-cylinder condensing turbine) Secheron generator	Speed	3,000 rpm	
		Pressure	31.4 bars	31 atm
		Temperature at inlet	360° C	680° F
		Power	Air-cooled, nominal power 7,750 kW Effective power 6,200 kW, 2,400 V, 50 Hz	

*Discontinued.

Level of Automation

The Cheneviers plant is highly automated. It can be run entirely from a central, glass-enclosed control room. The operators have a clear view of and easy access to all major components. A bank of system alarms will identify emergencies and sound a signal throughout the plant to alert personnel.

High pressure steam leaves the boiler and is fed to the high pressure distribution control group. From there, the steam goes to feed the main turbine, the feed water pump turbine, and as a reserve measure, to feed a low pressure bank. Steam is withdrawn from the turbine in four stages. The first passes through a reduction valve and provides the normal feed to the low pressure bank. The off-take from this bank is a deaeration system, combustion air preheaters, and services for the plant (hot water and heat). Second stage off-take of steam from the turbine feeds the second stage feed water reheater. Third stage off-take from the turbine heats the first stage feed water reheater. The final steam off-take from the turbine goes into the condensor.

The cooling water for the condensors is drawn from and redischarged into the Rhône River which runs next to the plant. The cooling water plant has a pumping and filtration system equipped with a Passavant screening system and 2 filtration beds, each capable of providing 3,600 m^3 (950,000 gal) of water per hour. There are also two Sulzer pumps each capable of delivering 1,750 m^3 (462,000 gal) per hour.

System Design Redundancy

There is a significant amount of redundancy in the system to insure that no vital element can bring the whole unit to a halt. The use of two parallel furnaces is the most obvious. This gives greater probability that steam will be available for the turbine and, more important, that waste can be burned at any time.

There is an auxiliary condenser allowing the furnace and boiler to function when the turbo-generator is not operating. There is a designed redundancy in the supply of steam to the low pressure bank, and also in the power systems for the boiler feed water pumps.

Within the steam cycle, all of the major pumps have been doubled, including the feedwater pumps for each boiler and the condensate pumps from the condenser. A second pump for condensate from the auxiliary condenser has been added because this auxiliary condenser works more often than originally anticipated.

The low pressure steam distribution system can be fed either with directly reduced steam from the high pressure bank or with first stage outlet steam from the turbine (see Figure 3.3).

FIGURE 3.3: BASIC STEAM CYCLE AT THE CHENEVIERS (GENEVA) INCINERATOR PLANT

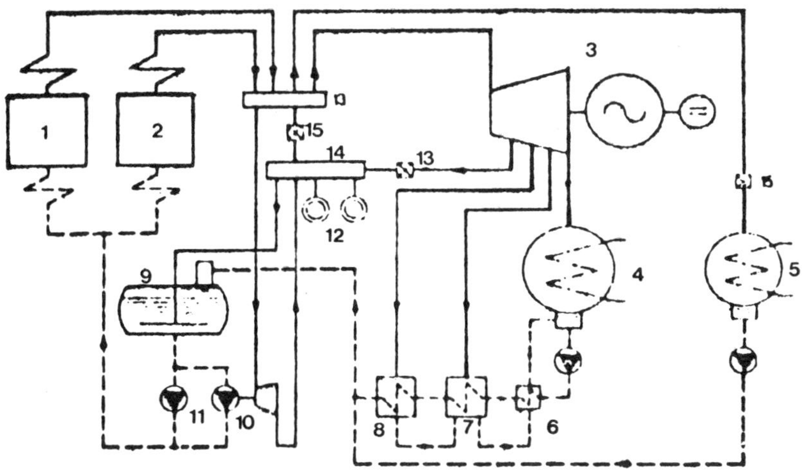

1. Furnace-boiler No. 1
2. Furnace-boiler No. 2
3. Turbo-generator
4. Main condenser
5. Auxiliary condenser
6. Supplementary cooling
7. First stage feedwater reheating
8. Second stage feedwater reheating
9. Feedwater storage with deaerator unit
10. Feedwater pump (steam turbine)
11. Feedwater pump (electric motor)
12. Internal steam consumption (low pressure)
13. High pressure steam valves
14. Low pressure steam valves
15. Steam pressure reducing valves

System Performance History

The total quantity of waste burned each year varies considerably, due less to fluctuations in quantities of waste than to the availability of the system to burn material. The amount of reduction in terms of ash per ton of burned waste has shown improvement during the last several years and now stands at nearly two-thirds reduction by weight. Table 3.5 shows a summary of the performance characteristics of the incinerators at Cheneviers.

Steam generation also varies depending upon the amount of time that the furnace is operating and the rate at which material is being burned. The performance figures at the bottom of this table show that about 2.4 metric tons of steam are recovered for each metric ton of refuse burned. This is superheated steam at 360°C (680°F) and 32.4 bars (32 atm).

When the furnaces are running, the average output of each furnace is from 15 to 17 metric tons (16.5 to 18.7 short tons) of steam per hour.

Electricity generation obviously also depends upon the performance of the plant and this has reached a maximum of 30 million kWh in 1975. In terms of

TABLE 3.5: SUMMARY PERFORMANCE CHARACTERISTICS OF THE CHENEVIERS INCINERATOR

SF/$	4.33	4.33	4.33	4.30	4.31	4.31	4.11	3.82	3.17	2.98	2.58
Characteristic	1966 6 mos.	1966/67 12 mos.	1967 6 mos.	1968	1969	1970	1971	1972	1973	1974	1975
Waste burned (metric tons x 10^3)	22.27	64.75	35.08	83.92	85.58	96.19	70.57	67.76	82.01	80.09	87.51
Ash produced (metric tons x 10^3)	8.17	27.50	14.21	39.29	38.03	41.58	28.25	23.72	29.66	29.74	31.13
Ash/waste (percentage)	36.70	42.50	40.50	46.80	44.50	43.20	40.00	35.00	36.20	37.13	35.60
Steam production (metric tons x 10^3)	29.27	147.80	84.15	189.50	197.40	220.30	181.90	169.40	197.80	192.10	208.20
Electricity											
Generated (kWh x 10^6)	2.19	18.60	8.71	23.95	25.43	30.86	24.56	23.60	25.60	28.84	30.93
Internally used (kWh x 10^6)	0.32	2.59	1.20	3.26	3.67	3.99	3.31	3.59	3.62	4.57	5.07
Bought in (kWh x 10^6)	1.22	1.00	0.60	0.69	0.46	0.36	0.62	0.48	0.93	0.44	0.39
Total used (kWh x 10^6)	1.55	3.59	1.79	3.93	4.13	4.35	3.93	4.07	4.55	5.01	5.45
Sold (kWh x 10^6)	1.87	16.38	7.84	21.43	22.67	26.50	20.63	19.53	21.04	23.84	25.48
Average sale price (ct/kWh)*	2.67	2.96	2.98	3.24	2.93	2.77	2.40	2.95	2.92	2.65	2.76
Value earned from electricity (Swiss Francs x 10^3)	42.60	495.20	234.90	694.70	665.10	735.20	494.20	576.30	614.90	632.40	704.20
Operational availability											
Incinerator 1 (percentage)		52.00	56.50	56.40	61.30	78.10	76.70	54.30	68.10	77.90	68.30
Incinerator 2 (percentage)		55.00	49.90	63.00	64.90	73.40	59.40	58.30	79.90	68.90	80.10
Turbine/generator (percentage)				75.00	84.70	91.50	78.90	83.80	75.70	89.10	89.10
Steam recovery (metric tons/metric ton waste)	1.31	2.28	2.40	2.26	2.31	2.29	2.58	2.50	2.41	2.40	2.38
Steam rate (metric tons/hour)		15.77	18.06	18.12	17.86	16.60	15.26	17.17	15.26	14.94	16.02
Electricity recovery (kWh/metric ton)	98.32	287.28	248.27	285.39	297.15	320.83	348.04	348.27	312.17	360.08	353.46
Waste burned (metric tons/day)	122.00	177.00	192.00	230.00	234.00	264.00	193.00	186.00	225.00	219.00	240.00

*Swiss Centimes per kilowatt hour.

electricity generated per ton of waste burned, this reached a peak of 360.08 kWh per metric ton (326.88 kWh per short ton) of waste in 1974.

The plant was designed to have a capacity of 200 metric tons (220 short tons) per day per furnace, but this was based upon waste with a 1500 kcal/kg (2700 Btu/lb) heat value. The actual waste apparently has considerably higher heat value and it has been found that the plant cannot function now above about a maximum of 170 metric tons (187 short tons) per day.

Materials and Energy Balance

There are only relatively simple data-gatherng and input-output analysis for the plant at Cheneviers. No detailed calculations of a materials or energy balance are made, nor is it felt that such analysis would be of much use because it would be expensive to acquire and complicated to calculate.

Furthermore, there is no guarantee that anyone would pay attention to the results. There is no regular analysis of the composition, heat value, or moisture content of the waste being treated. A thermal value test was made upon completion of the plant in 1966 and a small sample test subsequently attempted in 1969. There has been no such check since. There are obvious seasonal fluctuations in moisture content and, as was observed during the plant visit, some of the barges remain open and when it rains, the moisture content increases dramatically.

Maintenance History and System Modifications

The furnaces that Von Roll built for Geneva were of a relatively new design. Whereas earlier units had the boilers in a separate chamber, away from direct flames, and generally operated on saturated steam, here, the boilers were placed directly in the furnace as a waterwall. Superheaters were used for producing steam at higher temperature and pressure.

This new design seemed rational, both as a means of increasing the efficiency of energy recovery and as a means of reducing plant size. However, the lack of plant operating experience with this newer configuration of furnace and boiler meant that there were many unanticipated problems.

One of the first changes was a switch from the simple yearly shut down of each furnace for maintenance, as recommended by the manufacturer, to a program of cleaning and inspecting the furnaces and boilers every three months. These early inspections showed that major corrosion problems were developing. There were also problems of premature combustion on the drying grate and hot points on the main burning grate.

After several attempts to seek individual solutions, it was decided to make some major modifications to the system. To reduce the corrosion on the lower portion of the waterwall, a new air injection system was installed above the main combustion grate. The air generates turbulence in the gases and thus reduces the height of the flames which had been causing major damage to the waterwall tubes. The premature combustion on the drying grate was attributed to the higher calorific value of the waste and to the air stream that was injected beneath this grate in the initial design. This air supply was removed and the problem was solved.

However, an additional problem also occurred at the wall separating the drying and combustion grates where burning refuse accumulated causing a high concentration of heat. Here, a new injection system was installed to force air horizontally out through this wall. This caused the burning waste to move more rapidly down onto the burning grate.

The overall effect of these modifications was to increase the air flow through the furnace by about 20-25%. This increase necessitated greater capacity for gas extraction, and subsequently resulted in much more rapid wear in the induced draft fan and in the rate at which the cyclone units clogged up and had to be cleaned.

The furnaces originally had a special burner above the main grate for waste hydrocarbons. This unit never worked properly and was subsequently removed.

There were no major problems with the superheaters. One economizer has been removed and the dimensions of the others increased. An additional evaporator has been added.

The third passage in the furnace originally had a split route allowing the combustion gasses to pass either through the economizer or through an air preheater. This air preheater began to clog and was finally deemed unnecessary. It was removed during the recent overhaul of the plant. Inlet air is now preheated only by a steam-air heat exchanger to about 100°C (212°F). Previously, it had been preheated to 250°C (482°F).

The current routine maintenance schedule calls for each furnace to be stopped every two months for about a week for a cleaning of the critical points. Right now, the most vulnerable component in the system is the cyclone unit which, because of the higher air flow and the resultant decreased efficiency of the electrostatic precipitators, becomes rapidly clogged with dust. The plant manager feels that if this unit could be removed and replaced by larger electrostatic precipitators, the plant could probably operate for four to five months without shut-down.

Once a year, the furnaces need to be shut down for three to four weeks for major cleaning by sand blasting. A major overhaul (replacement of tubes and some refractory brickwork) is estimated to be required after about 30,000 hours of operation, or about every five to six years.

The plant originally had 10 inlet points in various parts of the furnace for steam lances to blow down and clean the tubes. These have been abandoned except for one location in each superheater. The steam lances are used once a day and operate on low pressure steam.

Overall Availability

The plant was originally expected to be in service 85-90% of the time, but this figure has never been attained. Table 3.6 shows the actual number of operating hours for each furnace since 1968, and the percentage of the total number of hours in the year that each unit has been available during that period. The individual performances vary from year to year due to unforeseen stoppages and to major repair work which can keep a furnace out of commission for three to four months.

TABLE 3.6: AVAILABILITY OF THE FURNACES IN THE CHENEVIERS INCINERATOR

	Furnace Boiler 1		Furnace Boiler 2		
Year	Total Hours	Percent of Possible Hours	Total Hours	Percent of Possible Hours	Average Percent
1968	4,939	56.38	5,520	63.01	59.69
1969	5,374	61.34	6,003	68.52	64.93
1970	6,844	78.12	6,019	68.71	73.41
1971	6,716	76.67	3,688	42.10	59.38
1972	4,758	54.31	5,453	62.24	58.27
1973	5,968	68.13	6,996	79.86	73.99
1974	6,830	77.97	6,040	68.95	73.46
1975	5,980	68.26	7,012	80.05	74.15

The two furnaces have been available about 75% of the time on the average during the last several years. In the opinion of the plant manager and the supervisors, this is about the best that can be expected from this plant. The availability of the turbo-generator has been much lighter and there have been no major problems with this unit.

Characteristics of Recovered Energy

Energy is recovered in the Cheneviers plant in the form of steam and electricity. The steam is produced only for internal use, for such system components as the air-preheaters, the deaeration tanks, etc. It also provides heat and hot water for personnel services.

Electricity is generated at 2,400 volts and 50 Hz. The turbo-generator operates on superheated steam at 360°C (680°F) and 32.4 bars (31.9 atm). It can operate on steam from either or both boilers, and thus can generate electricity whenever one furnace is operating.

The electricity produced is used in the plant and also sold to the grid of the local hydro-electric generating station at Verbois. When the Cheneviers plant is not generating electricity, this grid supplies plant requirements. The incinerator account is credited at the end of each year on the basis of net supply of electricity to the grid.

Table 3.5 shows how the steam and electricity output has varied from year to year in terms of quantity of waste incinerated. As there have been no systematic tests of the caloric value or moisture content of the waste, no cause can be ascribed to these variations.

The Market for Electricity

The Canton of Geneva's electricity is obtained from private electrical generation plants feeding into a publicly owned distribution grid. This grid is operated on a commercial basis, and can negotiate rates for obtaining its electricity independently, but is subject to controls on the charges that can be made for the sale of electricity.

The Cheneviers plant originally contracted to supply power at a rate far above its average output for ten hours of daytime and far below its average during the night. In actual practice, since its principal purpose is to eliminate waste, the plant has had to operate continuously and thus the electricity output is nearly uniform 24 hours a day.

According to the original contract, the owner of the distribution grid (the Services Industriels), promised to pay for the electricity at a rate that varied by month, following the schedule shown in Table 3.7. However, in view of the altered delivery schedule, the initial rates were reduced in 1970, and again in 1971 to a level where they remain to date.

Environmental Control Systems

When the Cheneviers plant was designed and built, there were few federal or cantonal environmental standards that had to be respected. However, the canton decided to limit particulate emissions to 160 mg/m^3, and installed a double bank of electrostatic precipitators followed by a cyclone. Since the installation of this plant, new standards have been adopted which call for a limit of 100 mg/m^3 for particulate matter. The plant has ten years to reach compliance with this new standard. The height of the stack had to be limited because of the proximity of the Geneva airport.

There are still no standards for emission of gases such as HCl or SO_2, although there are some constraints on the sulfur and chlorine content of fuels that can be burned. There has been no attempt to measure these elements in the waste that is burned in Geneva.

TABLE 3.7: SALE PRICE OF ELECTRICITY BY CHENEVIERS (Swiss Centimes per kilowatt hour)

Month	1966-69	1970	1971-present
January	4.5	4.2	4.0
February	4.5	4.2	4.0
March	4.0	3.5	3.5
April	3.5	3.0	3.0
May	2.5	2.5	2.0
June	0.7	1.0	1.0
July	0.5	0.5	0.5
August	0.7	1.0	1.0
September	2.5	2.5	2.0
October	3.5	3.0	3.0
November	4.0	3.5	3.5
December	4.5	4.2	4.0
ct/$.01	4.32	4.31	4.11-2.50

NOTE: Electricity used to be supplied to Cheneviers at a standard tariff rate when the plant was not functioning. However, with the revision in the sale price of electricity to the grid, it was agreed that electricity supplied to Cheneviers would be subtracted from electricity produced by Cheneviers and the plant would be credited with the difference at the above monthly rates.

There are no standards for the water used in the plant either. Water for condenser cooling and ash quenching is simply drawn from the Rhône and dumped back into the Rhône. There is a purification system for preparation of boiler feedwater which consists essentially of some settling tanks.

Internal working conditions in the plant are controlled by a variety of regulations. Material and operating standards are set by the Caisse National Suisse D'Assurance. This is a coordinating body for Swiss insurance companies that pressures policyholders to maintain proper health and safety standards. In 1963, when the plant was built, there were only a few standards. Today, there are many more.

Electrical standards are set by the Association Suisse des Electriciens (the Swiss Association of Electricians), while the Swiss Boiler Code comes from the Association Suisse des Proprietaires de Chaudieres (the Swiss Boiler Owners Association).

In general, the Swiss are slow, observant and thorough in the adoption of standards for the environment and for worker health and safety. There is a tendency to rely on the experience of West Germany and many of the current Swiss standards first appeared there several years earlier.

Aesthetics and Site

As mentioned above, when this plant was designed, money was readily available and a very modernistic and attractive plant was built. For aesthetic reasons, a single stack, rather than two, was built.

For practical reasons, it was decided to bring most of the waste to the plant by barge rather than by truck. In the first place, the plant is 16 kilometers (10 miles) from the center of town and this would have meant about 100 truck deliveries per day to the plant over city and cantonal roads. Second, the dead time during the back-and-forth trips wasted the truck capacities and the time of the crews. Thus, at considerable expense, a special dock was built near the middle of the city for the loading of barges from the collection vehicles.

The plant stands by itself on a bend in the Rhône River just upstream from the hydro-electric power plant at Verbois. There is still considerable land available for expansion, and the nearest town is small and some distance from the plant.

Comments from the Manufacturer of the Plant

Discussion with Von Roll revealed that this plant was one of the first to be built in the new design configuration. Older plants, upon which Von Roll had built its reputation, were equipped with refractory furnaces and tail-end boilers generating saturated steam. Energy was wasted and plants were unnecessarily large in this configuration. They anticipated savings by integrating the boiler into the firebox, recovering superheated steam, and reducing the size of the total unit.

Cheneviers and many other plants were built with this new design before the acute problems of corrosion were discovered. These first became evident in 1968. Subsequently, modifications were made to reduce corrosion and to make the plants more reliable. Refractory ceramic coatings were added to the lower portions of the waterwall tubes and special refractory material was added on the frontal walls where the flames had been found to be much higher than expected. A secondary air source was added above the burning grate to complete the combustion of volatilized matter and prevent the alternation of oxidization and reduction conditions in the chamber which was corroding the metal tubes.

Von Roll has since gone back to its original design of refractory furnaces with tail-end boilers, and has developed a variation on this design that allows the recovery of superheated steam. It also offers a redesigned version of the integrated boiler like the ones in Geneva, which has all of the superheater elements as well as the evaporators and economizers in the third pass of the furnace. A waterwall is still used, but the rest of the recovery is much farther downstream.

The firm also offers a new furnace with waterwalls and a horizontal passage for the combustion gases with superheated steam (a partial variation on the tail-end boiler theme). This boiler configuration allows the use of a mechanical rapping system for cleaning the tube packages. Von Roll feels that this system is superior to the steam blowdowns in their older furnaces.

ECONOMICS AND MANAGEMENT

System Management and Operational Control

The Cheneviers facility is designed to be managed and operated entirely at the plant. The building incorporates office space for the plant manager and his staff which consists of a secretary and an assistant. The manager is a qualified engineer who has worked at Cheneviers since it started operating in 1966.

The control room is designed so that a small crew can operate the plant. Instrumentation and automation permit full practical control of the system from this one central location. The crane is operated from an enclosed station and there are television cameras so that the crane operator can see into each location for picking up and unloading.

The data generated in the control room and at the loading scales could be used to analyze the operations on a much more detailed basis, but this is not done, and not felt to be necessary. Much of the information recorded is used only when an emergency arises and relevant data is needed.

Operating Crews

The plant originally had five-man operating crews with a relatively small maintenance staff on duty during the day. This situation has since changed. There are now five four-man operating crews which work around the clock on a five week schedule. The team composition, qualifications, and approximate salary level are shown below:

OPERATING TEAM COMPOSITION AND COST

Function	Qualification	Gross Salary SF
Foreman	CFC	52,000
Furnace Mechanic	CFC	44,000
Assistant Mechanic	No Special Qualification	40,500
Crane Operator	No Special Qualification	40,500
		177,000 ($70,800; 1976)

The CFC is a Certificat Fédéral de Capacité (A Federal Certificate of Aptitude) which is a non-university diploma. Each team is expected to work a 43.75 hour week. During the last seven years, total salaries have risen by 111%.

Seventeen men work on maintenance at the plant. This is a considerable increase over the force that had originally been anticipated. In addition, when major repair work is needed, teams of workmen from the central repair group are called in. The costs for these extra men in 1975 were the equivalent of five man-years. The maintenance team is composed as follows:

Team Member	Qualification
Foreman	CFC
2 Mechanics	CFC
2 Metal Workers	CFC
2 Electricians	CFC
1 Storekeeper	No Special Qualification
1 Painter	No Special Qualification
1 Greaser	No Special Qualification
7 Unskilled Workmen	No Special Qualification

There is also a scale operator and another workman during the day to regulate and record the arrival of the trucks. The contract with Von Roll called for the training of personnel before and during construction, and for Von Roll to guarantee the plant during the first six months of operation.

Original Cost and Amortization

The incinerator at Cheneviers went into service in 1966. The original cost was as follows:

	Swiss Francs	
Land Purchase	982,130.00	
Building & External Construction	13,529,605.60	
Equipment/Machines	14,590,193.80	
Total	29,101,929.40	($ 6,736,500; 1965)

In addition, the river barge transfer station at Cheneviers cost:

	Swiss Francs	
Land Purchase	--	
Building & External Construction	3,146,059.05	
Equipment/Barges	3,891,758.25	
Total	7,037,817.30	($ 1,629,100; 1965)

And, the dock installation cost:

	Swiss Francs	
Land Purchase	--	
Building & External Construction	3,661,834.60	
Equipment	672,186.25	
Total	4,334,020.85	($ 1,003,250; 1965)
and: Land Fill Alterations	101,232.20	($ 23,400; 1965)
for a Grand Total of	40,574,999.75	($ 9,392,360; 1965) or ($23,161,470; 1976*)

*Costs have been converted to U.S. dollars for the year in which the expenditure was made and then inflated to 1976 dollars according to the *Engineering News Record* Construction Cost Index.

Financing Terms

Land, building, and equipment financing were provided directly by the Treasury of the Canton of Geneva at a flat fixed rate of 4% per year. The loan for civil engineering work and building construction is being repaid over a 50 year period at a constant rate per year including the interest payment. This comes to a total of 4.655% per year.

The loan for the electromechanical installations is being paid back over 20 years at a fixed annual payment, including interest charges which comes to 7.358% of these costs. Land charges pay only the 4% interest per year. Thus the amortization per year is as follows:

Unit	Rate*	Swiss Francs
Cheneviers Incinerator		
Land	4%	39,285.20
Construction	4.655%	629,803.14
Equipment	7.358%	1,073,546.46
Total		1,742,634.80
Cheneviers Port		
Construction	4.655%	146,449.05
Equipment	7.358%	286,355.57
Total		432,804.62

(continued)

Unit	Rate*	Swiss Francs
Loading Dock		
Construction	4.655%	170,458.40
Equipment	7.358%	49,459.46
Total		219,917.86
Landfill Charges	4.655%	4,712.36
Grand Total		2,400,069.64 ($555,570; 1965)

*Four percent interest plus capital amortization.

Operating Costs and Revenues

Since the charges to the municipalities in the Canton of Geneva are based on the aggregation of all costs and operating revenues of the several plants, it is not possible to present a historical time series for the Cheneviers plant alone. However, a disaggregation in detail of costs and revenues has been made since 1973, and a copy of this breakdown for 1974 was obtained and is shown in Table 3.8.

As far as it has been possible to determine, these are real and actual cost and revenue items on the basis of which it is possible to have a very good idea of the operational components of this plant. The river barge facilities and the Cantonal landfill expenses are not included. These figures are only for the facilities at the Cheneviers plant.

As is obvious from the table, by far, the most important item in operating costs is labor. Labor represented 69% of total direct operating costs in 1974. One item in these labor charges needs to be mentioned specifically, the charge for Other Administrative Personnel. This item covers the cost of the staff which coordinates all of the treatment plants and should probably not be included in the actual operating costs of this one plant. It is a shared administrative overhead.

The other major item meriting attention is the cost of spare parts which is considerable and which represents replacement parts from a major rebuilding of the plant this year. In addition to these costs, there is an item at the bottom of the page which refers to a contribution to a special fund for spare parts. This fund was opened several years ago when it was realized that expenses were going to be much greater than originally anticipated. This fund serves as a sump for financing major overhaul work and allows such expenses to be spread over several years.

The Cheneviers plant's only revenue is from the sale of electricity. The expenses of this revenue came to 33SF per metric ton ($10.07 per short ton) of waste incinerated. Adding the contribution to the special account for spare parts, the cost came to 41.64SF per metric ton ($12.70 per short ton).

TABLE 3.8: OPERATING COSTS FOR THE CHENEVIERS INCINERATOR IN 1974

Operating Expenses	Swiss Francs	Dollars
Operating personnel	1,711,542.65	
Repair personnel	224,045.35	
Repair personnel-vehicle drivers	28,068.80	
Administrative personnel at incinerator	154,588.65	
Other administrative personnel	142,892.45	
Temporary personnel	1,718.80	
Subtotal personnel	**2,262,856.70**	
Work clothes	4,908.20	
Building maintenance (General Services)	16,562.25	
Maintenance/amortization of vehicles	60,498.80	
Laundry	940.65	
Supplies for central repair group	15,158.30	
Supplies for central stores	64,981.90	
Miscellaneous supplies	8,658.75	
Subtotal direct expenses	**171,708.85**	
Water supplies	80,579.05	
Telephone	4,749.10	
Spare parts	618,430.50	
Maintenance products	25,075.25	
Building maintenance	36,984.80	
Insurance	42,403.00	
Miscellaneous	37,465.25	
General account	760.50	
Subtotal other expenses	**846,447.45**	
Total expenses	**3,281,013.00**	**1,101,000**
Receipts from Sale of Recovered Materials		
Sale of electricity	632,445.00	
Expenses net of recovered product sale	2,648,568.00	
Special account for spare parts	686,866.00	
Total expenses	3,335,434.00	
Interest and amortization	1,810,823.20	
Total overall expenses	**5,146,257.20**	**1,726,900**
Total tonnage incinerated (tons)	**80,093.05**	
Cost per ton (without amortization)	**41.64**	**12.70/short ton**
Cost per ton (with amortization)	**64.25**	**19.60/short ton**

The amortization shown on this table is slightly smaller than that calculated earlier in this report. The discrepancy is due to the fact that an initial estimate of annual amortization charges was made before the final total costs were known. When it was revealed that actual amortization charges were slightly less, it was decided to continue with the original charge and to put the surplus in a fund to reimburse interest and amortization payments which were not charged during part of the first year, and also to cover the costs of the enlargement of the plant now under way.

The cost per metric ton treated including the amortization for the plant, came to 64.25SF ($19.60 per short ton). There are no known hidden charges or other subsidies in this statement.

During the same year, the costs per metric ton of material transported or treated at the different facilities in the Canton were as follows:

	COST PER METRIC TON (1974)**	
	Without Amortization	With Amortization
	SF	SF
Cheneviers	41.64	64.25
Richelien Incinerator	102.73	139.68
Villette Compost Plant*	66.55	80.01
River Transport***	15.62	28.26
Landfill	2.66	2.74

*Not charging this plant for some byproducts that had to be incinerated.
**$1.00 = 2.98 SF (1974)
***Including the rebuilding of one barge.

From the above comparison, it is understandable that there is pressure to close the incineration plant at Richelien. Until recently, the compost plant sent most of its low grade output directly to the landfill. The compost is now finding some favor for soil conditioning in the vineyards and the plant will be continued. For purposes of overall comparison, the consolidated cost and revenue statement for the Canton of Geneva in 1975 has been translated and is shown below in Table 3.9.

TABLE 3.9: CONSOLIDATED INCOME STATEMENT FOR ALL SOLID WASTE TREATMENT CENTERS IN THE CANTON OF GENEVA, 1975

Operating Expenses	Swiss Francs*	Percentage
Personnel and administration	4,331,285.60	49
Special clothing	7,515.75	–
Electricity	54,814.20	1
Water	65,673.95	1
Telephone	7,786.60	–
Maintenance equipment	1,104,192.00	12
Maintenance consumables/fuel	142,417.70	2
Building maintenance	92,721.90	1
Transport and vehicles	84,793.75	1
Insurance	55,334.70	1
Miscellaneous	87,778.35	1
Total	6,034,314.50	68
Amortization	2,800,000.00	32
Total expenses	**8,834,314.50**	**100**
Receipts		
Sale of electricity	704,210.00	
Sale of compost	46,519.30	
Sale of scrap iron	3,539.05	
Total recovery	**754,268.35**	

(continued)

Table 3.9: (continued)

Operating Expenses	Swiss Francs*	Percentage
Industrial disposal fees		
Ordinary waste	1,757,036.75	
Oversized waste	83,232.70	
Confidential waste	21,582.00	
Used hydrocarbons	7,374.60	
Dried sludge	408,224.60	
Filter waste from sewage plants	33,018.75	
Dangerous residues	99,409.60	
Non-putrifying waste	281,353.20	
Sand from sewage plants	17,398.95	
Liquid waste	43,006.35	
Total income	3,505,905.85	
Participation in interim interest charges for new Cheneviers plant	-(365,183.00)	
Total net income	3,140,722.85	
Remainder to be paid by municipalities	5,693,591.65	
At 88,857.23 ton = cost per metric ton	64.08	
Participation in fund for parts	4.92	
Total cost per metric ton for municipalities	69.00	

*In 1975, 2.46 SF = $1.00.

Plans for the Future

In 1970, four years after the incinerator at Cheneviers began functioning, a new study of the total solid waste situation in the Canton of Geneva was undertaken. This time a much more detailed analysis was made, covering all industrial and household waste (solid and liquid) with the aim of discovering how best to eliminate all nonsewer waste in a comprehensive system. There was growing pressure to reduce the amount of material in the cantonal dump to a minimum because space was severely limited.

The canton contracted with the Société Générale Pour L'Industrie (SGI) to carry out this study which cost approximately $85,000. The terms of reference were to forecast the amount of waste that would be produced during the next 20 years and to propose a system for its elimination that would minimize dumping and promote resource recovery. The only solution excluded in advance was composting, as the plant which was already operating was having unsatisfactory results.

The SGI presented its final report in 1972. Having considered traditional and modern treatment methods, SGI concluded that an extension of the incineration facilities should be built to handle the expected increase in regular solid waste. A new multipurpose incinerator was recommended to handle the various special categories of waste that were also to be treated at this plant in the pursuit of an absolute minimum of dumping. There was one additional constraint, however, on the eventual choice of a system. While new methods were desired and encouraged, no system could be considered which had not been in operation for 20 years. Thus, modernity was a relative concept.

The cantonal authorities asked the SGI to continue as consulting engineers for the new project, and made an initial request for proposals for a turnkey plant meeting a series of performance standards. They required the main contractor to provide not only the traditional furnace with recovery of energy (electricity) but also the multipurpose furnace for the special solid waste disposal. The results of the RFP were disappointing. Four major competitors, Von Roll (the only Swiss competitor and the supplier of the original Cheneviers incinerator), Martin (of West Germany), Bartolomeus (of Italy) and CEC (of France) submitted proposals which were all adequate for the traditional incinerator, but which were all unacceptable for the full range of performances required for the special furnace.

The canton and the SGI had to redefine their criteria and pursue a new design for the special furnace. This finally resulted in a choice of a rotary furnace from Kraus Maffei of West Germany. The contract for the traditional incinerator was awarded to Martin of Munich. Considerable political pressure was brought to bear to obtain the contract for Von Roll, but the cantonal authorities and SGI maintained that the technical design of the Martin plant was superior and that the price was better.

This plant is now being built adjacent to the original plant, and will be completed in late 1978. Martin will be the general contractor and will also be responsible for the installation of the Kraus Maffei furnace (with its related boiler, and other auxiliary systems). The new Martin furnace will have a capacity of 21 metric tons (23 short tons) per hour, while the rotary Kraus Maffei furnace will have a capacity of 4-5 metric tons (4.4-5.5 short tons) per hour for normal waste and less for special waste. The new combination will supply vapor to a BBC (Brown Boveri) turbo-generator set (which is about 33% more powerful than the original unit). All turbo-generator units will be interconnected so that steam flow can be optimized from the two older incinerators as well as from the two new units.

Financing for the new plant will come partly from subsidies from the federal and cantonal government, and partly from a bank loan to be negotiated with a cantonal guarantee for reimbursement.

According to the officials responsible, the procedures used in planning the extension of the plant were totally different from those used for the first plant. At that time, they had no experience and little ability to define exactly what was desired. As a result, the supplier gave them a standard system which, while good for its day, was in no way specially adapted for the particular requirements of Geneva. The extension on the contrary, was planned from start to finish by the authorities in Geneva with the help of a consulting engineering firm. This time there were many more specific ideas about what was needed and about how the plant should be delivered, guaranteed, etc. The canton tried to force the bidders to produce more imaginative solutions but also imposed prior operating history constraints which left little room for experimentation.

The financial situation is now very different from that when the first plant was built. The canton does not have a surplus of funds and cannot provide the financing for the new plant directly. Since a commercial bank will provide part of the funds, the loan will cost more and there will be other variations in the financing terms, duration, etc. The new plant will also have to meet more severe standards for pollution control and worker health and safety.

As in the case of the first plant, there was no consideration of an incinerator without the generation of electricity. It has not been possible to determine why no cost comparisons were ever made to determine the cost of recovering energy in this way. Energy recovery seems to be a political imperative that is not negotiable. It apparently makes no difference that it is a very expensive way to generate watts.

It is interesting also to note that while Von Roll was not selected to build the extension, Martin was not the first choice of everyone concerned. Those most closely connected with the present incinerator, the plant operator and supervisor, would have personally preferred to continue with Von Roll in spite of their past problems with corrosion.

CONCLUSIONS

The Cheneviers incinerator in Geneva shows the experience that a city has had with the first of a new generation of energy-recovery incinerators. The plant was selected relying largely upon the know-how and reputation of the supplier, and there appears to have been no analysis of the specific heat content of the municipal waste until the plant was delivered.

When this plant was ordered in 1962, it represented a new departure in waste incineration. There was an integrated boiler in the firebox to increase greatly the amount of heat that would be recovered. This design also allowed for production of superheated steam which can be used to generate electricity much more efficiently.

Cantonal Law provides for the operation of waste disposal facilities at cost (including amortization), but there is no ceiling and the municipalities must pay whatever it costs. While energy recovery generates some income for the canton, the operators consider that their principal concern is to reduce the volume of waste. Therefore, the most important factor is neither cost nor efficient recovery of energy, but continuous operation of the plant. The plant is therefore run at the level which is most likely to keep it in operating condition, regardless of the marginal tradeoffs of heat recovery efficiency.

For example, much thicker tubes than were originally specified are now used for the waterwall. This may have penalties for heat recovery, but will allow longer periods of operation before tube replacements are necessary. Modifications in the airflow, increased about 20% since first construction, have led to problems in the air filtering and evacuation end of the plant. The induced draft fans are wearing out more quickly now and so are the cyclone units. These latter are also clogging up so that furnaces must be shut down every two months for cleaning the cyclones.

Von Roll no longer offers this specific design. It has recognized the problems of corrosion and clogging that resulted and has now modified its integrated boiler units significantly. Von Roll also has other designs, recalling its earlier units with tail-end boilers, which it has been selling recently with success.

The actual cost of generating electricity at the Cheneviers plant cannot be determined. Since there is no way of comparing the existing operation to direct incineration without recovery of energy, it is not known whether this method of generating electricity is economically rational. The only available answer suggests that energy recovery is a political imperative. There are some who suspect that the generation of electricity in this way is inordinately expensive.

The plant was originally rated at 200 metric tons (220 short tons) per day per unit or 400 metric tons (440 short tons) per day total. This would give a maximum theoretical capacity of 146,000 metric tons (160,800 short tons) of waste per year. However, the plant has been available for only about 75% of the time, on average, in the last few years and thus its real capacity is only 300 metric tons (330 short tons) per day or 109,500 metric tons (120,600 short tons) per year.

Furthermore, the furnaces cannot be operated at 200 metric tons (220 short tons) per day even when they are working, since the waste is actually higher in calorie content than originally thought and the plant must, therefore, run at less than full capacity. This brings each unit down to about 170 metric tons (187 short tons) per day on line, which given the 75% availability, reduces the effective annual capacity to only 93,075 metric tons (102,500 short tons). This quantity was only exceeded once, in 1970, when 96,000 metric tons (105,700 short tons) were burned.

In 1975, the plant recovered 2.38 metric tons of superheated steam and 353.45 kWh of electricity for every metric ton of waste burned. The average operating rate was 16 metric tons (17.6 short tons) of steam per hour on-line.

REFERENCES

Literature

The following works were found useful as sources of information on the Cheneviers plant in Geneva.

(1) *L'Evacuation et le Traitement des Ordures Ménagères de L'Agglomération* Genevoise, Départment des Travaus Publics, République et Canton de Genève, May 1962.
(2) *Rapport d'Exploitation, Installations de Transport et de Traitement des Résidus,* Direction du Génie Civil, Service de l'Assainissement, Départment des Travaux Publics, République et Canton de Genève, for each year since 1966.
3) *Agrandissement des Installations Contonales de Traitment des Résidus, Rapport Technico-Economique.* Produced for the Division Entretien et Exploitation, Direction du Génie Civil, Départment des Travaux Publics, République et Canton de Genève, by the Société Générale pour L'Industrie, Geneva, March 1972. (Main report and ten annexes.)
(4) *Usine d'Incinération des Ordures Ménagères de Genève,* Sonderabdruck aus Schweizerische Zeitschrift für Hydrolgie, Birkhäuser Verlag, Basel, Vol. 13, 1969, Fasc. 2.
(5) Special section of the *Neue Zürcher Zeitung,* Wednesday, 1 June, 1966 describing the Cheneviers incinerator.
(6) "Les Déchets, Source de Matière Première," speech given by Mr. J.E. Chevallier, Chief of the Division of Operations and Maintenance, Public Works Department, Canton of Geneva, on 18 October, 1973.
(7) *Les Cheneviers, Usine de Destruction des Résidus de Genève,* special pamphlet describing the incinerator, prepared by the Department of Public Works for the inauguration of the plant, 2 June, 1976.
(8) Von Roll, Environment Engineering Division, an explanatory brochure provided by Von Roll, Zurich, Switzerland.

Persons Contacted

The following individuals and organization have cooperated in providing information about the municipal incinerator installation in Geneva.

Mr. Chevallier,	Chief of the Division of Operations and Maintenance, Department of Public Works, Canton Geneva.
Mr. F. Lancoud,	Chief of Sanitation Services, reporting to Mr. Chevallier.
Mr. Erne,	Plant Director of the Cheneviers incinerator, reporting to Mr. Lancoud.
Mr. Cousin,	Head of the Research Project undertaken by the Société Générale Pour L'Industrie, for Mr. Chevallier for the expansion of the Cheneviers plant.
The Von Roll Company	Manufacturer of the furnaces at Cheneviers.

Case Study of Korsor, Denmark

Material for this chapter has been based upon a report prepared by Resource Planning Associates, Inc. (RPA), Washington, D.C. and Paris (CONS/2103-3).

SUMMARY

The incinerator of Korsor is part of a thermal complex composed of an incinerator with heat recovery equipment and an oil-fired heating facility. Its originality is that it serves a small town of only 20,000 persons.

Incineration was chosen as the method of treatment because of the municipality's view that it is environmentally the best way to dispose of waste. District heating is very common in Denmark, so the decision to recover the heat for that purpose was not surprising. The fact that a Danish company, Bruun & Sorensen, specializes in small incinerators with heat recovery and has already sold many similar plants was certainly an essential factor in the decision. Construction of the incinerator was begun in October 1970, with Bruun & Sorensen as the main contractor, and was completed in May 1972.

The furnace has two combustion chambers and a separate cyclone-shaped postcombustion chamber. Bruun & Sorensen feels that, for small plants, this design allows for more complete combustion of gases and will consequently reduce corrosion.

The grate (an original Bruun & Sorensen design) has two sections with six horizontal axles apiece and grate prongs screwed onto the axles. A hydraulic system advances and agitates the waste by rotating axles 1, 3, and 5 in each section clockwise and axles 2, 4, and 6 counterclockwise. Heat is recovered by a fire tube boiler which produces steam at a maximum temperature of 120°C (248°F) and a pressure of 6 bars (5.9 atm).

The furnace has a rated capacity of 2 metric tons (2.2 short tons) per hour or 7,000 metric tons (7,700 short tons) per year for 3,500 hours of operation. Estimates of the amount of waste currently treated vary between 6,000 and 7,000 metric tons (6,600 and 7,700 short tons) per year. There are two limitations on the amount of waste treated: first, the amount of waste generated in the municipality and, second, the lack of equipment to dissipate the heat which cannot be sold during the nonheating period.

The design of the incinerator was based on the assumption that Korsor's waste would have an average lower heating value of 2,000 kcal/kg (3,608 Btu/lb). The actual lower heating value is estimated at 2,200 kcal/kg (3,968 Btu/lb). Household waste makes up about 70% of the total; the rest is essentially commercial waste and a little industrial waste.

The energy recovered from the boiler amounts to about 7,000 kilotherms (kth) per year. The thermal efficiency of the plant is estimated at approximately 50%. Taking into account energy which is lost because it cannot be sold, the overall efficiency decreases to approximately 30%.

The hot water produced is pumped into a district network serving a total of 1,350 apartments and houses, but no industrial plants at the moment. The main network is 7 km (4.3 mi) long; the overall length, counting the branches to individual houses, is 13.5 km (8.3 mi).

During the four years of operation, no major problem has occurred. The most serious problem has concerned the refractory material in the combustion zone. The refractory deteriorated very quickly and had to be changed within the first two years. However, the problem now appears to be solved.

To control pollution, the gases first pass through the postcombustion chamber where the ashes are partly precipitated, then (after the waste heat boiler) through a battery of cyclones. The supplier guaranteed the cyclone unit at 94% removal efficiency. No sampling of particulate emissions has been made.

The architecture of the complex is interesting and careful attention was paid to color, shape and structure. The interior is comfortable and the amenities are excellent. Construction costs (not including equipment) amounted to 46% of the total cost of the entire plant.

The total work force for the complex is 6 people, including the plant director, two operating shifts of two persons and one person for housekeeping. The plant is highly automated: one operator controls the crane operations, the furnace operations, and the heating facility from a central control room.

The total cost of the complex was 7,000,000 Kr ($944,700; 1971 or $1,437,600; 1976) with 4,600,000 of that sum going for the incinerator. There was no subsidy from the Danish government; the plant was entirely financed with municipal funds.

The complex is operated by the municipality as part of the Incineration and District Heating Works. The overall cost of incineration is estimated today at 120 Kr per metric ton ($18 per short ton) including both capital and operating cost, after making provision for the value of heat recovered.

The cost per ton could be reduced if there were more waste to treat. There is room for a second incinerator, but the complex would have to burn the waste of surrounding municipalities. Moreover, the extension of the district heating network has not been as fast as expected. Since there are no heat exchangers, dissipating the heat during summer months would be a problem unless new markets, especially in the industrial sector, can be found.

The heat is sold for approximately 92 Kr/kth ($3.79/$10^6$ Btu) (1975/1976 prices) which is lower than the market price of individual electric or fossil fuel heating. Surprisingly, the municipality did not try to estimate in advance the cost of incineration or the value of the recovered heat.

INTRODUCTION

Economic and Demographic Data

The town of Korsor is located in the extreme western part of Seeland, the main island of Denmark, about 100 km (62 mi) from Copenhagen. Korsor is a small seaport town of 20,000 people. There are two main industries: the State Railways Company and a glass company. The State Railways Company, which operates the ferries between the central islands, employs 2,000 people there, mainly for its nationwide catering activities. Scanglas operates a float glass plant which employs 800 people.

The town of Korsor has an area of 81.7 km^2 (31.5 sq mi) with an average population density of 245 inhabitants/km^2 (635 inhabitants/sq mi). The municipality is divided by a fjord into two districts, each with its own downtown center, linked by a bridge. The downtown centers are composed of stores and offices in one to three story buildings, while the suburbs are composed mainly of single family houses with a few apartment complexes. The Korsor incinerator is located at the southeastern edge of town, in the suburb of Tyreengen.

The town government consists of a borgmester (currently Mr. P. Hvidtfeldt Larsen) and a town council. The tax revenues of the town are 57,000,000 Kr ($9,421,000; 1976) for 1976 or 2,850 Kr ($471; 1976) per inhabitant, which is average for Denmark (1). A rough estimate of the average income of a working person is 60,000 Kr ($9,900; 1976) per year; very often a family will have two incomes adding up to 100,000 Kr ($16,500; 1976) per year (2). As is normal in Denmark, there is relatively little difference between after-tax incomes.

Waste Collection and Treatment

Waste collection in Korsor, as throughout Denmark, is the responsibility of the technical services of the municipality. Household waste is deposited in 100 liter (3.5 cu ft) paper bags positioned on metal stands outside the houses. When full, the bags weigh between 16 and 18 kg (35 and 40 lbs) (3).

The bags are collected once a week. The town uses two simple open trucks of 6 and 8 metric ton (6.6 and 8.8 short ton) capacity, with a crew of three, including the driver, on each. Waste is collected from 7,400 residences all year long and from an additional 830 summer residences for part of the year. In a few areas at the outskirts of Korsor, waste is collected by a private company.

Industrial and commercial waste is delivered by its generators to the municipality for treatment. Industrial waste is mainly packaging material, such as plastics from the catering industry. In January 1977, Korsor joined the centralized treatment system of Kommune Kemi A/S which treats oil and chemical waste from the whole country.

The amount of household waste collected by the municipality in 1975 was estimated at 5,200 metric tons (5,730 short tons) per year. This is equal to 0.7 kg (1.5 lb) per inhabitant per day. A further 2,500 metric tons (2,740 short tons) of industrial and commercial waste was brought to the municipality by private companies for treatment (4) (see Table 4.1).

TABLE 4.1: QUANTITY OF WASTE IN KORSOR (estimated in metric tons*)

	1973	1974	1975
Supplied to Incineration Plant			
Domestic waste	4,523	3,056	4,268
Other combustible waste	2,954	3,056	1,759
Total	7,477	6,572	6,027
Dumped			
Domestic waste	517	524	891
Other combustible waste	1,087	671	842
Total	1,604	1,195	1,733
Total Collected			
Domestic waste	5,040	4,039	5,159
Other combustible waste	4,041	3,728	2,601
Total	**9,081**	**7,767**	**7,760**

* There is no scale at the plant. Number of truckloads obtained from municipal records. RPA estimates based on the following ratio: 2.2 metric tons (2.4 short tons) per truck for domestic refuse and 1.8 metric tons (2.0 short tons) per truck for other refuse (commercial and some industrial refuse). One metric ton equals 1.1 short tons.

The amount of household waste collected apparently varied from 5,040 metric tons (5,552 short tons) in 1973 to 4,039 metric tons (4,450 short tons) in 1974 and back again to 5,159 metric tons (5,682 short tons) in 1975. The quantity of industrial waste also decreased from 3,900 metric tons (4,300 short tons) in 1973 to 3,600 metric tons (3,960 short tons) in 1974, and decreased further to 2,500 metric tons (2,750 short tons) in 1975 (5). These figures are only estimates, as the municipality does not weigh the waste collected, but only records the number of truckloads. As we shall see, there are differing estimates which would substantially change the cost per ton.

Waste is treated in two ways, incineration and landfill. In 1975, approximately 83% by weight of the household waste (4,300 metric tons or 4,736 short tons) and 68% of the industrial and commercial waste (1,700 metric tons or 1,870 short tons) were incinerated (6).

The remaining waste (900 metric tons or 990 short tons of household waste and 800 metric tons or 880 short tons of industrial and commercial waste) was dumped in a sanitary landfill along with wood and other demolition waste. The residues of incineration are also dumped in this landfill.

The total cost of municipal collection amounts to 1,250,000 Kr ($204,250; 1975/1976) in 1975/76, or 240 Kr/metric ton ($35.60/short ton). This appears high for Denmark for a service which provides one collection per week (7). The cost of maintaining the landfill amounted to 120,000 Kr ($19,600) for 1975/76 or 15 Kr/metric ton ($2.20/short ton) of waste (household, industrial and commercial, and construction waste) (7)(8).

These expenses are covered by the sale of the collection bags by the municipality. The basic price for these bags is 224 Kr ($37; 1976) for 52 bags. Additional bags cost less. The basic price for 52 bags is expected to increase from 224 Kr ($37) to 285 Kr ($47) in 1977 to cover additional costs, in particular the purchase of a new truck (9). The sale of bags brought in 1,480,000 Kr ($241,830; 1975/1976) in 1975/76 and will contribute 2,000,000 Kr ($330,575) to the 1977 Korsor town budget. This covers the cost of collection and the cost of maintaining the landfill; there is no profit.

Revenues from the dumping of industrial waste amounted to 60,000 Kr ($9,800; 1975/1976) in 1975/76. Private companies pay 20 Kr ($3.27) per truckload brought to the landfill.

THE DEVELOPMENT OF THE SYSTEM

Reasons for New System

Before the construction of the incinerator in 1972, the waste generated in Korsor was simply dumped into the fjord, in an area more or less protected by a dam made of construction waste. This practice was abandoned as a detriment to the environment.

The choice of incineration with heat recovery was based on three considerations. First, at the time the decision was made in the 1960s, it was felt that incineration was the best way to protect the environment. Moreover, at that time, incineration, and especially incineration with heat recovery, was fashionable. This solution had been chosen by many other Danish municipalities. Finally the lack of landfill space was an argument in favor of incineration (10).

Other possibilities for waste treatment were apparently not seriously considered. Landfilling was eliminated because of the lack of suitable land for controlled dumping. At that time the municipality was not aware of the environmental quality standards possible in a good controlled landfill. Composting was also eliminated because of environmental considerations (11). District heating is very common in Denmark, and some towns already used heat recovered from incineration for district heating.

Project Development

In 1965, the housing complex of Tyreengen expressed its interest in a municipal district heating scheme. The Town Council began consideration of the construction of an incinerator in 1966, as the dumping of the waste in the fjord was becoming unacceptable. No study of alternatives was made, but members of the Council and the Technical Services visited several Danish incinerators and consulted with officials of other towns (12).

In 1967 the Council decided to study sites for an incinerator and district plant (13). Through discussion with other town officials, the members of the Korsor Town Council were convinced that the Danish company Bruun & Sorensen offered the best system for Korsor.

The municipality then asked Bruun & Sorensen and two other Danish incinerator manufacturers, Volund and E. Rasmussen, to submit bids. The amounts of all three bids were similar and the Town Council awarded the contract to Bruun & Sorensen (14).

The Bruun & Sorensen project was formally approved by the municipality at the end of 1969. There was no major opposition in the Council and some very enthusiastic support. There was no opposition at all from environmental groups or private citizens (15).

Because nobody on the Council or even in the Technical Services of the municipality was a specialist on incinerators, Bruun & Sorensen had to help the municipality in drawing up the specifications. Their advice was followed without hesitation. For example, for construction of the building, the Council took Bruun & Sorensen's advice and contracted with the Kjaehr & Trillingsgaard Company (16).

Construction began in October 1970. The district heating plant began operation in November 1971 and the incinerator in March 1972, after 17 months of construction. This seems a little longer than usual, but the municipality insisted that everything be completed before the incinerator start operation (17).

TECHNICAL INFORMATION

The Incinerator (18)

Waste is transported by trucks to a closed discharge hall where it is emptied into two bunkers of a total capacity of 560 m^3 (732 cu yd) (see Table 4.2 and Figure 4.1). One bunker is used for household waste and the other for commercial and industrial waste, so that it is possible to feed the furnace with waste of different calorific values, as needed. The bunker can store three to four days' worth of waste.

TABLE 4.2: KORSOR INCINERATION UNIT TECHNICAL DATA

Dates		
Start of construction	October 1970	
First district heating plant operation	November 1971	
First incineration of waste	March 1972	
Plant Dimensions	**Metric System**	**English Equivalent**
Dimensions of building	36 m x 26 m	118 ft x 85 ft
Height of discharge hall	.12 m	39 ft
Height of furnace room	18 m	59 ft
Height of heating plant	12 m	39 ft
Height of waste heat boiler	14 m	46 ft
Height of smokestack	62 m	203 ft
–diameter of first flue	1.1 m	3.6 ft
–diameter of second flue	0.8 m	2.6 ft
Incinerator		
Number of furnaces	1	
Rated capacity (per hour)	2 metric tons	2.2 short tons
Hot water production at normal capacity, lower heating value of 2,000 kcal/kg (3,607 Btu/lb)	3.0 kth/hr @ 6 bars, 120° C	1.2 x 10^7 Btu/hr, 248° F
Volume of waste pit	560 m^3	732 cu yds
Volume of crane bucket	1 m^3	1.3 cu yds
Gas produced	6 m^3/kg of waste	96 cu ft/lb of waste
Volume of gas in cyclone battery	22,000 m^3/hr	776,800 cu ft/hr
Efficiency of cyclone battery	94%	
Maximum quantity of particulates	300 mg/m^3	
Average temperature of gas	225° C	437° F
Yearly capacity (3,500 hours/year)	7,000 metric tons	7,700 short tons
Heat Recovery System		
Weight of fire tube boiler	25 metric tons	27.5 short tons
Input temperature of gas	900° C	1,652° F
Output temperature of gas	225° C	437° F
Water content of boiler	14 metric tons	15.4 short tons
Heat production	3 kth/hr	1.2 x 10^7 Btu/hr
Maximum water temperature	120° C	248° F
Working water pressure	6 bars	5.9 atm
Heating Facility		
1 oil fired boiler	5 kth/hr	2.0 x 10^7 Btu/hr
1 oil fired boiler	4 kth/hr	1.6 x 10^7 Btu/hr
1 oil fired boiler	2 kth/hr	7.9 x 10^6 Btu/hr
Length of main network	7 km	4.3 mi
Length of total network	13.5 km	8.4 mi
Number of housing units served	1,350	
Outlet temperature	85-110° C	185-230° F
Return temperature	60-70° C	140-158° F

FIGURE 4.1: SCHEMATIC DRAWING OF KORSOR INCINERATOR

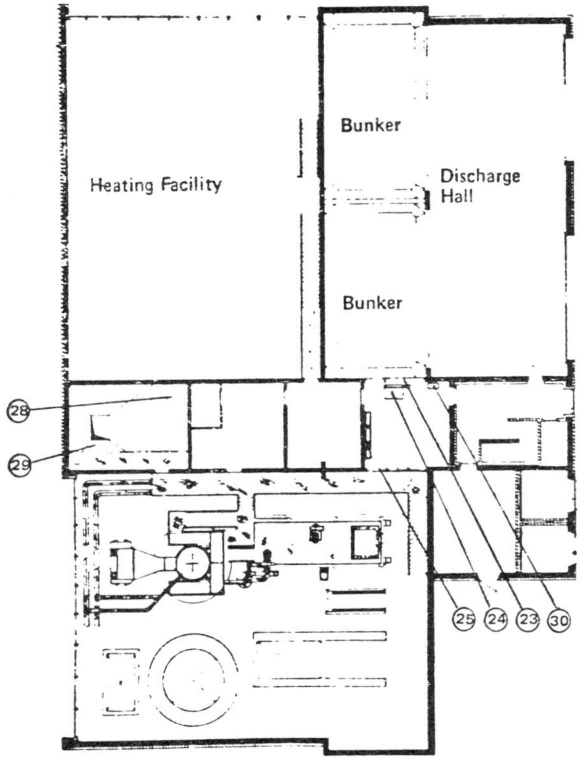

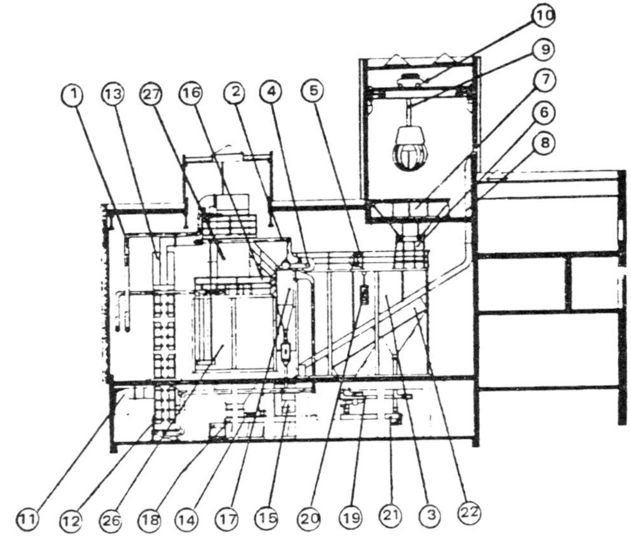

1. Tube arrangements
2. Gas return
3. Incinerator
4. Conveyor fan
5. Secondary fan
6. Waste chute
7. Waste hopper
8. Duct for combustion air
9. Crane with 1 m^3 bucket
10. TV camera
11. Gas duct
12. Exhaust fan
13. Dust collector
14. Primary fan
15. Slag discharge
16. Stairway
17. Secondary cyclone
18. Vibrating conveyor
19. Primary air duct
20. Gas burner (igniter)
21. Screw conveyor
22. Grate
23. TV monitor
24. Control desk
25. Control panel
26. Post combustion chamber
27. Boiler
28. Hoist
29. Slag silo
30. Panel for crane and hoist

A three metric ton (3.3 short ton) moveable crane with a one cubic meter (1.3 cu yd) bucket transports the waste from the discharge bunkers to the furnace hopper. This semiautomatic crane can feed the furnace approximately 16 times an hour. Waste is introduced into the furnace through an inlet hopper, a furnace shaft, and a hydraulic feeding door with refractory lining. The furnace room is designed to accommodate two furnaces, but for the time being there is only one furnace with a capacity of 2 metric tons (2.2 short tons) per hour.

The furnace is built on the two chamber principle with a separately lined cyclone shaped post combustion chamber. Bruun & Sorensen generally use this system for their small municipal incinerators. The design ensures more complete combustion of the gaseous emissions, and consequently reduces corrosion.

The furnace itself consists of a steel supporting framework, a refractory lining, and a hydraulically operated grate. The side plates consist of 5 mm (0.2 in) steel; the supporting steel construction is made of sectional iron; the top of the furnace is covered with floor plates.

The grate is Bruun & Sorensen's W Type, an original design by Mr. Sven Alexandersson, manager of their environmental engineering department (see Figure 4.2). It has two sections with six horizontal axles in each section. The cast steel grate prongs are approximately 0.5 m (20 in) in length and 5 cm (2 in) in breadth and are screwed to the axles. Transport and agitation of the waste are ensured by a hydraulic system which causes axles no. 1, 3 and 5 in each section to turn clockwise, while no. 2, 4 and 6 turn counterclockwise.

Each grate section can work at a speed adapted to the type and quantity of waste to be burned. Each section rocks through 60°; the whole grate is inclined 28° to the horizontal.

The refractory lining at the center of the furnace is made of different types of charmotte bricks; just above the grate line there are wear resistant cast iron plates. The furnace is insulated with moler bricks and the exterior temperature remains at approximately 35°C (95°F).

The air for incineration is supplied by primary and secondary fans. The primary air is supplied to the individual combustion zones through a system of channels arriving under each grate section. This air is drawn from the discharge hall in order to reduce air pressure slightly and thus contain odors.

The flue gases from the incinerator are kept between 800°C (1470°F) and 1050°C (1920°F). The lower limit has been designed to produce odorless flue gases, the upper limit is the melting point of the ash. The ash goes from the last incineration grate into the ash shaft where a ram moves it onto a vibrating conveyor located under the furnace. Before the ash enters the vibrating conveyer, it is cooled and dust is reduced by a water spray. The ash is finally lifted by a hoist into a silo with an outlet at the bottom.

FIGURE 4.2: BRUUN & SORENSEN TYPE W GRATE

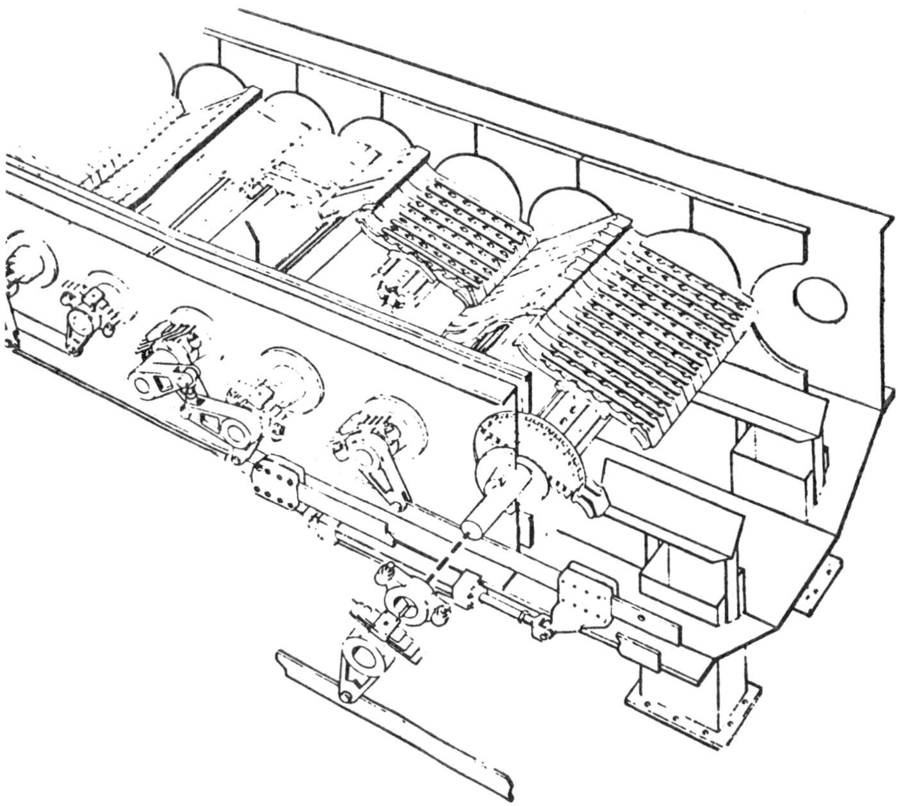

Heat Recovery Equipment (18)

The flue gases from the furnace pass into the postcombustion chamber, then into a fire tube boiler which recovers the thermal energy they contain. The input temperature of the flue gases is approximately 900°C (1650°F); they leave the boiler at approximately 250°C (480°F). The output of the waste heat boiler is 3 kcal (3.6×10^7 Btu) per hour at a working pressure of 6 bars (5.9 atm) and a maximum temperature of 120°C (240°F).

To prevent overheating in the boiler there is a sensing device which can automatically shut the feeding door. The boiler is cleaned with compressed air.

After the boiler, the gases pass through a battery of 26 cyclones; the fly ash separated in the cyclone battery is transported pneumatically to a secondary cyclone, and from there to the ash conveyor.

The exhaust fan placed after the cyclone units draws the gases up the stack. The stack is clean when the incinerator is operating.

Heating Facility (19)

The oil fired heating facility, located in the same building as the incinerator, also produces hot water for the district heating network. In addition to the heat recovered from incineration of waste, the heating facility burns oil to fire three boilers rated at 5, 4 and 2 kth/hr. There is room for an additional boiler in the facility. The different capacities of the three boilers allow them to operate at nearly full capacity, which results in greater efficiency. The water comes from the municipal distribution system and is treated in the plant.

The oil fired boilers are only used during the heating season. Water is heated to 85°C (185°F) and returns to 60°C (140°F). During the summer, the oil fired boilers are not used. However, because there is still more heat available from the heat recovery system than needed during the summer, steam is pumped into the network at a temperature of 110°C (230°F) and returns at 70°C (158°F).

The hot water is pressurized and pumped into a district heating network with a total length of 13.5 km (8.4 mi). The main network is 7 km (4.3 mi) long and there are 6.5 km (4.1 mi) of branches to houses. This network serves 1,350 apartments and houses, but no industrial plants at the moment.

Steam Efficiency

According to municipal records, the incinerator burned 6,000 metric tons (6,600 short tons) of waste in 1975. However, as pointed out previously, this figure has been disputed and some estimates are that 7,000 metric tons (7,700 short tons) were treated. The municipal records show that 4,300 metric tons (4,700 short tons) of household waste and 1,700 metric tons (1,870 short tons) of industrial and commercial waste were treated.

The furnace has a rated capacity of two metric tons (2.2 short tons) per hour, or 7,000 metric tons (7,700 short tons) a year for 3,500 hours of operation. In practice, it is possible to treat 2.3 to 2.5 metric tons (2.5 to 2.7 short tons) per hour (19). The plant operates 80 hours a week, with the incinerator operating an average of 66 hours a week (20).

The furnace was designed for waste with a lower heating value of 2,000 kcal/kg (3,608 Btu/lb), while the actual lower heating value is estimated at 2,200 kcal/kg (3,968 Btu/lb) (19) although no sample has been measured. The energy which can be recovered is estimated by the municipality at 1,300 kcal/kg (2,345 Btu/lb) of waste burned for a thermal efficiency of almost 60%. RPA estimates that this figure is probably too high. An efficiency of 50% seems more realistic.

In practice during the operating year 1975/76, a total of 7,046 kth (2.8×10^{10}

Btu) of heat was recovered from the incinerator, measured at the level of the waste heat boiler. (See Table 4.3.) On the basis of 6,000 metric tons (6,600 short tons) of waste burned per year and a lower heating value of 2,200 kcal/kg (3,968 Btu/lb), the overall efficiency rate of the plant would be 53%. If 7,000 metric tons (7,700 short tons) are treated per year, the rate would be only 46%.

TABLE 4.3: PRODUCTION AND SALE OF HEAT

	1973/74	1974/75	1975/76	1976/77
District Heating Facility Incinerator				
Oil consumption (metric tons*)	2,156	2,323	2,511	3,318
Calorific value (kcal/kg)	9,700†	9,700	9,700	9,700
Efficiency (percent)	87	87	87	87
Heat production (kth**)	18,196	19,604	20,707	28,000
Incinerator heat production	7,383	6,580	7,046	7,000
Total heat production (kth)	25,579	26,184	27,843	35,000
Heat sold (kth)	21,986	19,841	21,283	26,500
Losses (kth)	3,593	6,343	6,560	8,500
Total heat loss (percent)	14.0	24.2	23.5	24.3

SOURCE: Municipal sources–budget for 1977.

* To convert metric tons to short tons, multiply by 1.1.
** One kth (kilotherm) equals 3.97×10^6 Btu.
† 9,700 kcal/kg = 17,500 Btu/lb.

The heat obtained from the incinerator provides 25% of the total energy for district heating (1975/76). The remaining heat, or 20,707 kth (8.22×10^{10} Btu), was furnished by the oil fired heating facility. The thermal efficiency of the oil fired boilers in the heating facility is estimated at 87%.

Further inefficiencies, external to the incineration plant, are due to losses in the district heating network and the fact that not all available heat can be sold. During the summer there is no market for the heat produced.

The plant has no heat exchange system to dissipate the heat which cannot be used. Even when there is no need for district heating, the heat must be sent through the district heating network. During the summer months, the only energy needed in the houses is for hot water which is obtained through heat exchangers located in each house. In order to dissipate the heat which cannot be sold during these periods, the water temperature at the exit of the plant is raised to 110°C (230°F) (instead of 85°C [185°F] during the heating season) so that energy losses in the district heating network are maximized. The return temperature of the water is 70°C (158°F) during the nonheating season, instead of 60°C (140°F) during the heating season.

Thus the energy losses in the district heating network result in part from technical inefficiencies, but are partly voluntary. The losses in the district heating network amounted to 23.5% of the total energy sent into the network in

1975/76. They have varied between 14 and 24% from year to year, depending on the energy demands of the customers. It is estimated that a loss of roughly 10 to 15% represents the real technical inefficiency of the distribution network and that an additional 10% represents heat sacrificed because of the lack of a market (21). In 1975/76, losses due to lack of a market amounted to about 2,800 kth (1.11×10^{10} Btu) or from 18 to 21% of the energy available in the waste incinerated.

Overall, based on the 1975/76 figures and on the assumption of 6,000 metric tons (6,600 short tons) of waste treated, the efficiency of the energy transformation can be estimated at 53% in the incineration plant. If one takes into account the heat which is lost because it cannot be sold, this efficiency decreases to 32%. Based on the assumption of 7,000 metric tons (7,700 short tons) of waste treated, efficiency of energy transformation can be estimated at 46%; subtracting the heat lost because it cannot be sold, this efficiency would decrease to 28%.

System Reliability

The furnace was designed to operate 3,500 hours a year at a capacity of 7,000 metric tons (7,700 short tons) per year. In practice, during the year 1975, the furnace operated for 3,433 hours, or 98% of the available hours. Cleaning operations took a total of 252 hours and the furnace was not operating because of a breakdown for 20 hours (see Table 4.4) (22).

TABLE 4.4: FURNACE OPERATIONS, 1975

	Furnace in Operation (hours)	Breakdown (hours)	Cleaning (hours)
January	296	18	18
February	277	0	23
March	258	0	21
April	317	5	24
May	297	1	22
June	306	0	23
July	108	0	0
August	295	0	22
September	331	3	25
October	330	3	28
November	303	3	19
December	315	0	27
Total	**3,433**	**20**	**252**

SOURCE: Plant records.

All breakdowns have been very short. During 1975 they averaged only a few hours each month. The furnaces were cleaned on a regular monthly basis. The budget for maintenance was 150,000 Kr ($24,500; 1975/1976) in 1975/76 and 200,000 Kr ($30,050; 1976) in 1976/77.

During its four years of operation, no major problem has occurred at the incineration plant. The most significant problem concerned the refractory material in the combustion zone of the furnace. This deteriorated within the first year, and six square meters (65 sq ft) of refractory material were replaced under the guarantee. During the second year, a further six square meters (65 sq ft) had to be replaced. According to the manufacturer, the damage could be due to the high calorific value of the waste. According to the plant manager, it was due to bad workmanship (23). In any case, no further repair of the refractory has been necessary. It has also been necessary to replace the cast steel liners on each side of the grates; these liners deteriorate from the heat and need to be changed every three years.

A number of minor modifications were needed. A vibrator was put on the ash silo to shake the ash down and the cable on the crane and the cooling jacket of the screw conveyor were replaced.

According to the manager, the incineration unit is in good shape and should last for 25 years with the replacement of a few essential parts such as the boiler and the cyclones. It is, however, also his opinion that the heating facility is more reliable and of better quality than the incinerator.

Pollution Control

The volume of gas produced is 6 m^3/kg (96 cu ft/lb) of waste incinerated (24). After leaving the furnace, the gases pass through the postcombustion chamber and the waste heat boiler, then through multi-cyclone collectors before being evacuated by the exhaust fan through the stack. The ashes are removed partly in the postcombustion chamber and partly in the multi-cyclone collectors. From the postcombustion chamber, the fly ash is transported by screw conveyor to the ash conveyor of the plant. The fly ash separated in the cyclone battery is transported pneumatically to a secondary cyclone and, from there empties onto the ash conveyor.

The particulate content of the gases has never been tested, so it is necessary to rely on data from similar plants and on the manufacturer's guarantees. According to the manufacturer, the separation of fly ash in the postcombustion chamber is highly effective and plays an important role in preventing corrosion of the boiler. Another plant by the same manufacturer in Rosendall, Holland, has two furnaces, each having a capacity of 4 metric tons (4.4 short tons) per hour. At Rosendall, tests have shown that 60 kg (132 lb) per hour of fly ash are separated in the postcombustion chamber; 40 kg (99 lb) per hour of fly ash are separated by electrostatic precipitation. The remaining particle content of the exhaust gases is 100 mg/m^3 (25).

The Korsor plant is not equipped with an electrostatic precipitator, but only a battery of cyclones. According to the manufacturer, an electrostatic precipitator is much more efficient than a cyclone battery (on the order of 150 mg/m^3 after treatment with an electrostatic precipitator against 400 mg/m^3 with cyclone battery), but an electrostatic precipitator is more expensive. The manufacturer also claims that it is not necessary for a plant such as Korsor's. The municipality of Korsor confirmed this point by saying that in Denmark there is no regulation requiring any particular treatment or establishing a maximal level of dust emissions. There are some recommendations from the state environmental agency, but most incineration plants in Denmark do not have electrostatic precipitators at present. The manufacturer guarantees that the multi-cyclone collectors will provide a dust separation efficiency of 94%, with no particulates weighing more than 400 mg and a total particulate content no greater than 300 mg/m^3.

The stack has two flues, one for oil fired boilers (1.1 m or 3.6 ft in diameter) and one for the incinerator (0.8 m or 2.6 ft in diameter). The temperature of the flue gases at the exhaust is approximately 225°C (437°F). No analysis of the chemical content of the gases has been made.

The slag and ash amount to approximately 5% of the original waste by volume and 20-25% by weight. The manufacturer guarantees that unburned material will represent between 4 and 6%, by weight, of original waste, and that putrescibles will be less than 2.5% by weight, according to EAWAG method (*Eidgenossige Anstalt fur Wasserversorgung Abwasserreinigung und Gewasserschutz,* Zurich). The dumping of the residues in the landfill does not present any problem; nonincinerated waste is dumped along with the residues of incineration.

According to the municipality, there have been no complaints from the neighborhood or environmental groups. A large housing complex is located 100 m (110 yd) from the incineration plant and new dwellings are under construction within 30 m (33 yd) of the discharge hall.

At this distance, it is estimated that, without some protective barrier, odors and noise from the discharge hall will be perceived by the future inhabitants. When this was mentioned, representatives of the municipality said that the managers of the development are aware of this problem. The Bruun & Sorensen representative said that in another municipality, Alberstulund, the incinerator is located just as close to the mayor's office (26).

According to the municipality, environmental groups consider incineration favorable to the environment and will not take any action against it. The municipality held a two-day open house when the incinerator was inaugurated in order to show those interested how it operates.

Building and Working Conditions

The building is 36 m (118 ft) long and 26 m (85 ft) wide. The discharge hall is a closed building, 12 m (39 ft) high, and has two main doors. Behind the discharge hall, the two bunkers and the crane form another block 18 m (59 ft) high. Part of the furnace room is on the left, and the battery of cyclones occupies a third block 12 m (39 ft) high with a projection 14 m (46 ft) high for the waste heat boiler. The plant is located next to the fjord without any fence or screen of trees. The construction and civil engineering work was done by Kjaehr & Trillingsgaard under the supervision of Bruun & Sorensen.

The buildings were carefully designed both for external appearance and for interior comfort and amenities. Particularly liked were the structure, shape, and outside colors of the building which made an attractive contrast with most industrial plants. Working conditions appear very satisfactory. The plant is operated essentially from the glass enclosed control room which has a view of the silos and crane on one side and of the furnace on the other.

The plant is highly automated and is operated by a single person in the control room. The operator controls the furnace automatically and the crane semiautomatically. The crane moves automatically to precoded positions but the opening and closing operation must be performed by the operator. A TV camera mounted on the crane makes it possible to supervise the feeding operation itself. The control room also has time recorders which control all keyed functions, e.g., the start-up of the vibration conveyor and the ash hoist which run intermittently.

Visitors are impressed by the cleanliness of the working areas and especially of the machinery. According to the plant manager, cleaning operations are very important, not only to insure satisfactory working conditions, but also to insure proper maintenance of the machinery and reduce risk of accident.

No accident has ever occurred in the plant (27). At night, the heating facility works automatically during the heating season and the entire plant is linked to an alarm system by telephone. The alarm has gone off a few times but nothing serious has ever happened.

ECONOMICS AND MANAGEMENT

Management

Incineration of waste and production and distribution of heat are managed by the municipality of Korsor. The Chief Mechanical Engineer of the municipality is in charge of the Water Works, the Gas Works and the Incinerator and District Heating Works. He reports to the Town Council (see Figure 4.3). The director of the plant reports to the Chief Mechanical Engineer.

In addition to the director, the Incinerator and District Heating Works employs eight people. Two shifts of two people each carry out the operation of the plant and one person is responsible for housekeeping; in addition, three persons work at the maintenance of the district heating network. One of the members of each shift is in the control room operating the crane and the furnace; the other is on the floor and helps with cleaning and maintenance.

FIGURE 4.3: ORGANIZATION CHART FOR THE INCINERATOR AND DISTRICT HEATING FACILITY

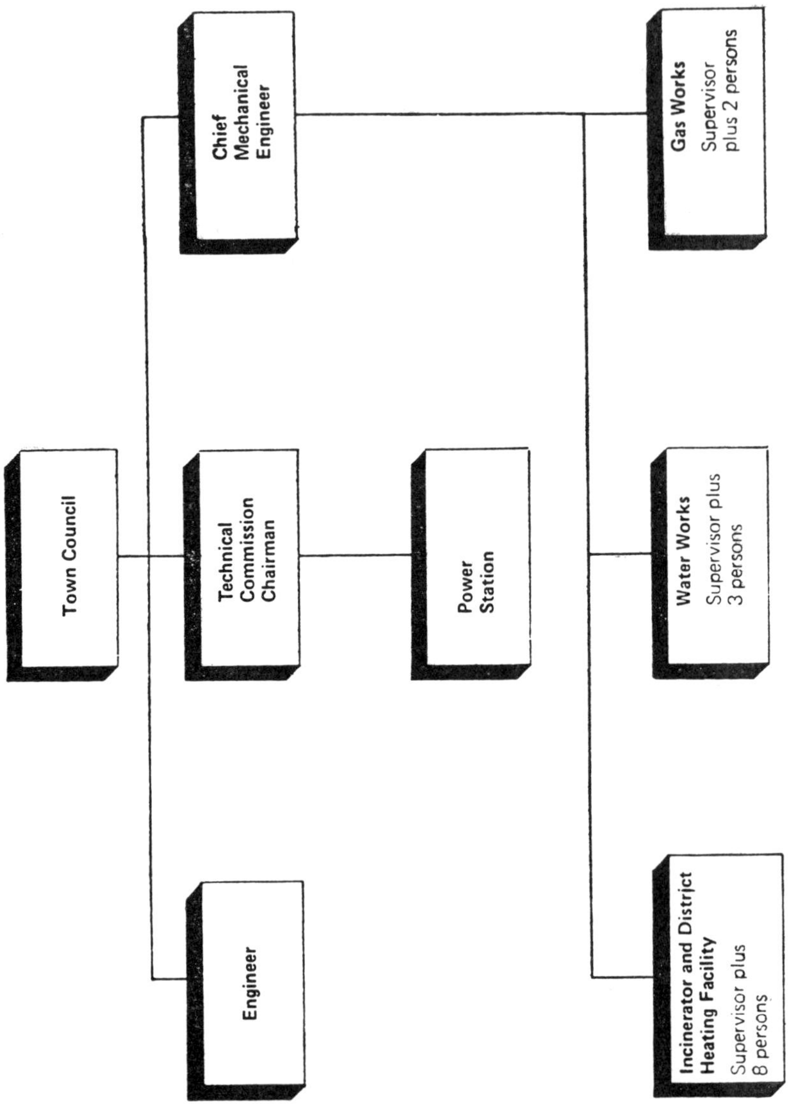

SOURCE Personal communication with Mr. Samer.

The manufacturer provided some training assistance when the plant opened. The workers at this plant come from other municipal departments. They are usually given a few weeks trial on the job before being permanently assigned to this department. Three to four months experience is required before the employee can control operations, particularly for the furnace. According to the plant director, the employees have only three to four hours of operational work per day; most of this time is spent in housekeeping.

The plant director has 8 years experience as a ship's engineer and 7 years experience with the boiler manufacturer, Bruun & Sorensen. The Chief Mechanical Engineer of Korsor also worked for Bruun & Sorensen.

The town of Korsor exercises supervision and control through budgets and financial reports. Apparently the town has never requested a technical report or cost evaluation of the incinerator or heat production facilities.

Capital Costs

The cost of the incinerator with the heat recovery system was 4,600,000 Kr ($620,800; 1971) and the cost of the heating facility was 2,400,000 Kr ($323,900; 1971) for a total cost of about 7,000,000 Kr ($944,700; 1971 or $1,437,600; 1976) (28) (see Table 4.5). The largest expenditures were made in 1970/71 (28% of the total cost) and in 1971/72 (55% of the total cost) as shown in Table 4.6.

TABLE 4.5: CAPITAL COSTS, 1966-1973

Incinerator		
Land	172,324.45 Kr	
Building	2,344,765.87 Kr	
Equipment	22,853.65 Kr	
Mechanical installation	1,699,861.60 Kr	
Design and other	382,071.57 Kr	
Total, Incinerator	**4,621,877.14 Kr**	
District Heating Facility		
Land	67,310.63 Kr	
Building	915,874.99 Kr	
Equipment	8,926.73 Kr	
Mechanical installation	1,303,624.69 Kr	
Design and other	149,238.70 Kr	
Total, District Heating Facility	**2,444,975.74 Kr**	
Total	**7,066,852.88 Kr**	**$953,700** (1971 dollars)

SOURCE: Municipal Report, Project Account for Incinerator District Heating Facility

Of the total cost of the incinerator, 50% went for construction and 37% for mechanical equipment. For the heating facility, construction costs were 37% of the total and mechanical equipment costs were 56%. The estimated value

TABLE 4.6: DATES OF EXPENDITURES BY THE MUNICIPALITY FOR THE INCINERATOR AND DISTRICT HEATING FACILITY

1966/67	2,969.12 Kr	
1967/68	2,811.33 Kr	
1968/69	14,675.10 Kr	
1969/70	493,283.70 Kr	
1970/71	1,935,367.73 Kr	
1971/72	3,916,504.73 Kr	
1972/73	659,241.17 Kr	
Reserve fund for major repairs	42,000.00 Kr	
Total	**7,066,852.88 Kr**	**$953,700** (1971 dollars)

SOURCE: Municipal Report, Project Account for Incinerator District Heating Facility.

of the land is 240,000 Kr ($32,400; 1971) or 3% of the total cost of the plant (see Table 4.5) (28). There was no subsidy from the national government. The plant was financed entirely by municipal funds; no loans were necessary (29).

For amortization purposes, the life expectancy of different components is assumed to be:

buildings:	33 years	3% amortization rate
equipment:	12.5 years	8% amortization rate
machinery:	12.5 years	8% amortization rate

Thus, the cost of amortization for the plant is 220,000 Kr ($36,360; 1976) per year (28).

Although there was no financial cost because the municipality used funds on hand, the municipality applies an opportunity cost of 9% to the portion of the cost which has not been yet amortized. This opportunity cost will be 315,000 Kr ($52,070; 1976) for the operating year 1976/77. Adding the opportunity cost to the cost of amortization gives a total capital cost pf 535,000 Kr ($88,430; 1976) to the operating year 1976/77.

Direct Operating Costs

The cost of operation of the incineration plant is reported by the municipality to be 943,164 Kr ($154,100) for the operating year 1975/76 (see Table 4.7) (29). This figure includes 77,000 Kr ($12,600) for administration of the Incinerator and District Heating Works (part of the salary of the Chief Mechanical Engineer and his secretary). The administrative offices are not located in the plant.

Salaries make up 57% of direct operating costs. The average salary of a worker at the plant is about 65,000 Kr ($10,620) per year, while the salary of the

TABLE 4.7: DIRECT OPERATING COSTS FOR THE INCINERATION PLANT, 1975/76

Salaries for operating personnel	464,654 Kr	
Salaries for administrative personnel (part of joint expenses)	63,000 Kr	
Pension fund	14,307 Kr	
Taxes and insurances	10,211 Kr	
Maintenance of building	15,044 Kr	
Maintenance of mechanical installation	152,784 Kr	
Electricity and water	70,178 Kr	
Office expenses· (part of joint expenses)	12,138 Kr	
Removal of ash	25,848 Kr	
Value added tax	115,000 Kr	
Total	**943,164 Kr**	**$154,100** (1975/76 dollars)

SOURCE: Data taken from municipal reports.

Chief Mechanical Engineer is about 150,000 Kr ($24,510) per year (29). Maintenance of the building and mechanical installations accounts for 18% of total operating costs and taxes represent 13% of total operating cost.

The direct operating costs according to the city's figures would be either 134 or 157 Kr per metric ton ($19.90 or $23.40 per short ton) incinerated, depending upon whether it was assumed that 6,000 or 7,000 metric tons (6,600 or 7,700 short tons) of waste were incinerated.

However, these official figures are too high, as the salary account covers all personnel working in the plant. According to the director, his employees spend 75% of their time working on the incinerator and 25% working on the heating facility. If, therefore, we allocate 75% of the salaries to the incinerator and 25% to the heating facility, we arrive at a total cost of 827,000 Kr ($136,000) or a cost per ton incinerated of between 118 and 138 Kr per metric ton ($17.50 or $20.40 per short ton), depending upon assumptions of tonnage.

The fees paid by private companies which bring their waste to the plant to be incinerated should be subtracted from this figure. In 1975/76, revenues from these fees were estimated at 20,275 Kr ($3,350) or about 20 Kr ($3.30) per truckload. This reduces the cost by approximately 3 Kr ($0.50) per ton incinerated.

A figure representing the value of the heat recovered from waste should also be subtracted. The municipality has never made an estimate of this amount.

In 1975/76 the town sold 21,283 kth of heat at a total price of 1,958,000 Kr ($320,100) or 92 Kr ($15.21) per kth. The amount of heat exported from the plant was 27,843 kth (1.1×10^{11} Btu) of which 7,046 kth (2.8×10^{10} Btu) were produced from waste. It is estimated that the operating cost of the heating facility alone amounts to roughly 1,410,000 Kr ($230,400) in 1975/76 of which 84% represents fuel costs.

It is also necessary to estimate the capital cost of the heating facility. The amortization rate is estimated at 93,000 Kr ($15,200; 1975/1976) based on a depreciation rate of 3% (33 years) for the building, 8% (12.5 years) for the equipment and 5% (20 years) for the mechanical installation. To this amortization rate has been added a financial cost of 9% per year of the portion of the cost which has not yet been amortized or 184,000 Kr ($30,065; 1975/1976).

It is estimated that the cost of producing approximately 21,000 kth (8.34×10^{10} Btu) in the heating plant is 1,687,000 Kr ($275,650; 1975/1976) or 80 Kr ($13.10) per kth. Consequently the value of the 7,000 kth recovered from waste in 1975/76 is estimated at 560,000 Kr ($91,500) or 80 to 93 Kr per metric ton ($11.90 to $13.80 per short ton) incinerated, again depending upon which assumption of quantity is adopted.

If the revenues received from private companies and the estimated figure for the value of the heat recovered are deducted from the direct operating cost of the incinerator, an operating cost of 246,725 Kr ($40,800; 1976) or 35 to 41 Kr per metric ton ($5.20 to $6.10 per short ton) in 1975/76 is arrived at, depending upon the assumption about the quantity of waste incinerated.

Overall Costs

The total cost of incineration to the municipality can be obtained by adding the capital cost of 535,000 Kr to the operating cost of 246,725 Kr. This gives a total cost in 1975/76 of 781,725 Kr. This equals 112 Kr per metric ton ($16.30 per short ton) based on the assumption of 7,000 metric tons treated or a total cost of 130 Kr per metric ton ($19.40 per short ton) based on the assumption of 6,000 metric tons treated. An average value would give a total cost of incineration of about 120 Kr per metric ton ($18 per short ton). Table 4.8 summarizes the overall cost.

The figure of 120 Kr per metric ton is only a rough estimate, as it was necessary for RPA to allocate joint costs between the incinerator and heating facility. This figure will decrease as the financial cost (which is an important element of the total cost) was obtained by applying a 9% opportunity cost to the portion of these costs which has not been depreciated.

Surprisingly the municipality has not tried to evaluate these costs. As a result, fees for the incineration of waste and charges for the sale of heat do not represent the true costs. For example, district heating, which is less expensive than individual gas or electric home heating, does not cover the cost of recovering the heat. Also the price that private companies pay for the treatment of their waste is far from covering the total cost of incineration or even the operating cost of the incinerator.

TABLE 4.8: OVERALL COST OF INCINERATION

	Kr	$ (1976)
Capital Costs		
Amortization	220,000	36,360
Opportunity Cost (9%)	315,000	52,070
	535,000	88,430
Operating Costs		
Direct Operating Cost	827,000	136,700
Less: Revenue from private generators of waste	20,275	3,350
Estimated value of heat from waste	560,000	92,550
	246,725	40,800
Total Cost		
	781,725	129,230
Cost per Ton (on the basis of 6,500 metric tons per year)	120 Kr/metric ton	$18/short ton

SOURCE: Data taken from municipal reports.

REFERENCES

Notes

(1) Personal communication, Mr. Samer.
(2) Personal communication, Mr. Alex Boisen.
(3) Personal communication, Mr. Samer.
(4) Municipal records.
(5) RPA estimate based on municipal records.
(6) ibid.
(7) Municipal budget.
(8) The cost of maintaining the landfill rises to 70 Kr per ton if construction waste is excluded.
(9) Personal communication, Mr. Samer.
(10) Personal communication, Mr. Jensen, Vice-Mayor of Korsor, Member of the Council and Chairman of the Technical Committee.
(11) Personal communication, Mr. Samer.
(12) Personal communication, Mr. Jensen and Mr. Samer.
(13) Municipal reports.
(14) Personal communication, Mr. Samer.
(15) Personal communication, Mr. Jensen.
(16) Personal communication, Mr. Samer.
(17) Personal communication, Mr. Boisen.
(18) All data provided by the manufacturer.
(19) Data provided by Mr. Riegels Jensen, Plant Manager.

(20) Data taken from plant records.
(21) Personal communication, Mr. Riegels Jensen, Mr. Boisen.
(22) Plant records.
(23) Personal communication, Mr. Riegels Jensen.
(24) Data supplied by the manufacturer.
(25) Personal communication.
(26) Personal communication, Mr. Samer and Mr. Boisen. A visit to the Albertslund incinerator confirmed this.
(27) Personal communication, Mr. Riegels Jensen.
(28) Municipal Report—Project Account for Incineration—District Heating Plant.
(29) Municipal records.

Persons Contacted

The following persons have cooperated with RPA in preparing this case study:

Mr. Alexandersson	Manager, Waste Treatment Department, Bruun & Sorensen
Mr. Boisen	Engineer, Bruun & Sorensen
Mr. Knudsen	Engineer, Bruun & Sorensen
Mr. Samer	Director, Incinerator and District Heating Works, Municipality of Korsor
Mr. Jensen	Technical Advisor, Council of the Municipality, Municipality of Korsor
Mr. Riegels Jensen	Plant Manager, Incinerator and District Heating Works, Municipality of Korsor

Case Study of Munich, Germany

Material for this chapter has been based upon a report prepared by Resource Planning Associates, Inc., Washington, DC, and Consultex, SA, Geneva, Switzerland (CONS/2103-4).

SUMMARY

The City of Munich, capital of the West German Lande of Bavaria, has a population of 1,315,000. Solid waste is collected by a city department and delivered to the city's own Electricity Works. The Electricity Works incinerates the waste in five incinerators in two separate locations (the North and South plants).

Munich's five incinerators represent three different designs for cogeneration of heat and electrical energy. All burn both waste and fossil fuel and all use Benson boilers to produce steam at 184 bars (181.5 atm) and 540°C (1000°F). The Electricity Works recovers energy in the form of electricity, hot water, and steam. Ash, bulky objects, and nonincinerable waste are dumped on a "garbage mountain" within the city limits.

The first design (North IA and IB) has two separate fireboxes (one burning waste, the other coal), but a shared boiler in each incinerator. The second design (North II) burns both waste and coal in one firebox for one boiler. The third design (South IV and V) has separate fireboxes for waste and fossil fuel, and separate boilers. The waste fired boiler in this arrangement is used as an economizer for the fossil-fuel-fired main boiler. All the plants use Martin grates and Babcock boilers. The designs were developed by the suppliers and city engineers working with planners from the Electricity Works.

The system's priority is to produce electricity for the city grid rather than to treat waste. The hot water and steam produced are used in district heating. The system is peak-loaded both for electricity and for district heating, and, as a result, is operated with some diseconomy.

Over 450,000 metric tons (495,700 short tons) of waste were burned in 1975. That amount represents practically all the solid waste collected in Munich (only bulky objects are not incinerated). A shredder with a capacity of 30,000 metric tons (33,000 short tons) per year was added in 1977 and this allows for even more complete incineration. Technical data on all the units are shown in Tables 5.1, 5.2 and 5.3.

TABLE 5.1: MUNICH NORTH IA AND IB UNITS TECHNICAL DATA
(Designed characteristics per unit)

	Metric System	English Equivalent
Daily waste burning capacity	600 metric tons	661 short tons
Hourly waste burning capacity	25 metric tons	27.5 short tons
Heat content of waste	900-1800 kcal/kg	1619-3238 Btu/lb
Guaranteed performance	25 metric tph at 1500 kcal/kg	27.5 short tons per hour at 2700 Btu/lb
Grate area	56.1 m^2	603.6 sq ft
Grate width	6.7 m (three grate runs)	22 ft
Steps per grate	17 (8 fixed, 9 moving)	
Maximum steam rate	100 metric tons per hour	110 short tons
Maximum pressure	205 bars	202.3 atm
Working pressure	184 bars	181.4 atm
Fresh steam temperature	540°C	1000°F
Fresh steam pressure	182 bars	179.6 atm
Feed water temperature	250°C	482°F
Flue gas particle content after the electrostatic precipitator	30 mg/m^3 at 8% CO_2	
Turbine unit maximum continuous rating	68 MW	
Turbine normal operating rating	62 MW	
Turbine steam inlet pressure	181 bars	180.7 atm
Designed heat from waste	40%	
Designed heat from fossil fuel	60%	
Flue gas temperature		
—coal only	120°C	248°F
—coal and waste	140°C	284°F
Boiler efficiency		
—coal only	92%	
—coal and waste	83%	
—waste only	72%	
Fuel characteristics: coal		
—heat content	7000 kcal/kg	12,600 Btu/lb
—ash	5%	
—water	8%	
Fuel characteristics: municipal waste		
—heat content	1500 kcal/kg	2700 Btu/lb
—ash	20-25%	
—water	28-33%	

TABLE 5.2: MUNICH NORTH II INCINERATOR* TECHNICAL DATA

	Metric System	English Equivalent
Daily waste burning capacity	960 metric tons	1050 short tons
Hourly waste burning capacity	40 metric tons	44 short tons
Heat content of waste	900-2000 kcal/kg	1619-3600 Btu/lb
Grate area	100.6 m^2	1082.4 sq ft
Grate width	11.3 m (five runs)	37 ft
Steps per grate	17 (8 fixed, 9 moving)	
Maximum steam rate	365 metric tons per hour	402 short tons
Maximum pressure	205 bars	202.3 atm
Working pressure	184 bars	181.5 atm
Live steam temperature	540°C	1000°F
Live steam pressure	182 bars	179.6 atm
Feed water temperature	266°C	511°F
Flue gas particulate matter content after electrostatic precipitation	18 mg/m^3 at 8% CO_2	
Turbine unit maximum continuous rating	112 MW	
Turbine normal operating rating		
Turbine steam inlet pressure	181 bars	180.7 atm
Designed maximum heat from waste	20%	
Designed heat from fossil fuel	80-100%	
Flue gas temperature		
–waste only	178°C	352°F
–waste/coal	165°C	329°F
Boiler efficiency		
–coal and waste	89.7%	
–waste only	81.4%	
Furnace efficiency		
–combined firing	98.3%	
–waste only	94.1%	
Unburned matter in ash		
–combined firing	1.65%	
–waste only	1.51%	
Organic matter in ash		
–combined guarantee	0.3%	
–combined tested	0.064%	
–waste only guarantee	0.3%	
–waste only tested	0.025%	
Fuel characteristics: coal		
–ash	6.56%	
–heat content	7155 kcal/kg	12,900 Btu/lb
Fuel characteristics: waste		
–ash	35%	
–heat content	1530 kcal/kg	2750 Btu/lb

*The characteristics of the North II incinerator were defined by tests performed upon acceptance in May 1968. Tests were conducted by TÜV Bayern.

TABLE 5.3: MUNICH SOUTH IV AND V UNITS TECHNICAL DATA

(Designed characteristics per unit)

	Metric System	English Equivalent
Daily waste burning capacity	960 metric tons	1050 short tons
Hourly waste burning capacity	40 metric tons	44 short tons
Heat content of waste	900-2000 kcal/kg	1619-3600 Btu/lb
Guaranteed performance	40 metric tph at 1500 kcal/kg	44 short tons per hour at 2700 Btu/lb
Grate area	101 m^2	1087 sq ft
Grate width	11.3 m (5 runs)	37 ft
Steps per grate	17 (8 fixed, 9 moving)	
Maximum steam rate	365 metric tph	402 short tons per hour
Maximum pressure	205 bars	202.3 atm
Working pressure	181 bars	180.7 atm
Live steam temperature	535°C	995°F
Live steam pressure	181 bars	180.7 atm
Feed water temperature	260°C	500°F
Flue gas particulate content after the electrostatic precipitator	105 mg/m^3 at 7% CO_2	
Turbine unit maximum continuous rating	124 MW	
Turbine normal operating rating	110 MW	
Turbine steam inlet pressure	180 bars	179.7 atm
Designed heat from waste	20%	
Designed heat from fossil fuels	80-100%	
Flue gas temperature		
–natural gas only	120°C	248°F
–gas and waste (mixed)	175°C	347°F
Boiler efficiency		
–gas only	93.5%	
–gas and waste	84%	
Fuel characteristics: waste		
–heat content	1500 kcal/kg	2700 Btu/lb
–water content	30%	
–ash content	34%	
Fuel characteristics: gas		
–heat content	8600 kcal/m^3	967 Btu/cu ft

Waste is collected once a week, and both private and industrial and commercial generators pay the city an all-inclusive fee based on volume. Part of this fee goes to pay for collection and part goes towards the difference between the total costs of incineration and the revenue earned from the sale of energy and scrap metal. The rates charged appear to be far below actual costs of incineration, which suggests that the sale of electricity is subsidizing the cost of incineration.

The entire North plant (IA, IB, and II) cost 151.1 million DM ($37.8 million; 1960-1966 or $88.3 million; 1976). The waste treatment section is estimated to have cost 25.8 million DM ($6.45 million; 1960-1966 or $10.2 million; 1976). The entire South plant (IV and V) cost 140.7 million DM ($39.2 million; 1969-1971 or $67.9 million; 1976). The waste treatment section cost 35 million DM

($9.8 million; 1969-1971 or $13.5 million; 1976). Official figures show the average cost per metric ton of waste treated (after deductions for the sale of the energy recovered) to be 18.75 DM ($6.92 per short ton) at the North plant and 18.50 DM ($6.83 per short ton) at the South plant. It is felt that these operating costs are serious underestimations, and that the correct figures are approximately 53 DM per metric ton ($19.56 per short ton) at the North plant and 41 DM per metric ton ($15.13 per short ton) at the South plant.

None of the designs has proved entirely satisfactory and for its new solid waste treatment facilities, Munich will be building incinerators which burn waste only.

INTRODUCTION

Economic and Demographic Data

The city of Munich comprises most of the area of Landkreis München and is the capital of the Lande of Bavaria, the extreme southeastern state of the Federal Republic of Germany. The city of Munich covers a roughly oblong area of 310 square kilometers (120 square miles), has a perimeter 117 kilometers (73 miles) in length, and is situated at an average altitude of 530 meters (1,739 ft) above sea level. The city is generally flat and in its eastern part, the Isar River flows through it. The two incinerator plants are situated on this river, about 20 kilometers (12.4 miles) apart.

The population of the city of Munich at the end of 1975 was 1,315,000. The city contains 546,500 dwellings within 101,200 residential buildings. There are a couple of neighboring communities for which domestic solid waste is also treated. This adds an additional several thousand persons to the total served by the city plants. The population density of the city of Munich is 4,242 inhabitants per square kilometer (10,986 per square mile).

The total gross volume of business of the city of Munich in 1974 was 84.0 billion DM ($32.4 billion). This represents a per capita turnover of 64,000 DM ($25,500). Average incomes are relatively high compared to most other countries in Europe, and total family incomes tend to be even higher due to a large number of working wives.

There are about one thousand manufacturing companies in Munich with 750 plants of various sizes. However, there is little heavy industry. Much of the activity is confined to office and administrative work, light industry, handicrafts, tourism, and the production of some very good beer.

Waste Collection and Treatment

Figure 5.1 shows graphically how the generation of incinerable solid waste has evolved in Munich since 1965. The volume increased from 960,000 m^3

FIGURE 5.1: QUANTITY OF SOLID WASTE COLLECTED AND BURNED SINCE 1965

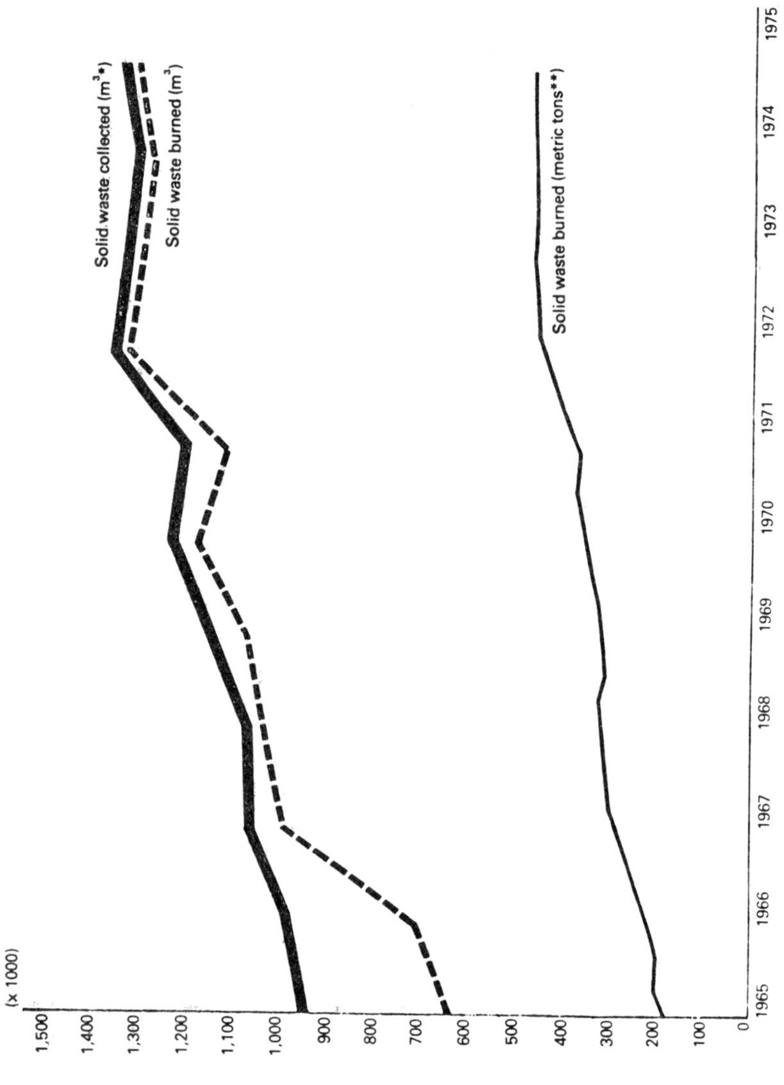

*To convert cubic meters to cubic yards, multiply by 1.3
**To convert metric tons to short tons, multiply by 1.1

(1,255,000 cu yd) in 1965 to 1,340,000 m^3 (1,724,000 cu yd) in 1975. By weight, incinerable solid waste grew from 180,000 metric tons (198,300 short tons) in 1965 to about 450,000 metric tons (495,700 short tons) in 1975.

This incinerable waste came from households and other private generators (commercial and industrial) in the following amounts:

Source of Incinerable Solid Wastes

Year	Households	(metric tons) Other Private	Total
1970	296,368	51,618	347,986
1971	297,861	48,134	345,995
1972	339,781	92,208	431,989
1973	359,716	84,172	443,888
1974	373,586	69,331	442,917
1975	377,626	72,501	450,127

There are no facilities at this time for the burning of oversized incinerable waste. This is estimated to represent about 40-50,000 metric tons (44-55,000 short tons) per year which are at present being dumped.

Large sized and nonburnable solid waste amounted to 480,000 m^3 (627,800 cu yd) in 1975. The corresponding figures were estimated to have been 407,000 m^3 (532,300 cu yd) in 1973 and 493,000 m^3 (644,800 cu yd) in 1974.

Household waste is collected by a special department of the city of Munich, the Städtische Müllbeseitigung. There are 150 packer trucks (MAN/Ochsner) having a capacity of 10 m^3 (13 cu yd) or 4-5 metric tons (4.4-5.5 short tons) of waste each. The collection department employs 655, including 213 drivers and 38 vehicle repair personnel. The work week is 42 hours.

Collections are made from Monday morning through Friday afternoon. All of the waste collected is delivered to the two refuse incineration plants (Munich North and Munich South) where the trucks empty their loads into storage pits. Lack of storage capacity at the Munich South plant forces suspension of deliveries of waste to this plant early in the day on Friday.

All other waste, including oversized burnable waste, is hauled to a controlled dump (a veritable garbage mountain). Incinerator ash is also hauled from the two plants to this dump. Almost no burnable waste (other than large size pieces) is now dumped.

The Städtische Müllbeseitigung charges private and public generators of wastes an overall collection and treatment charge on the basis of the volume of waste collected. This price includes a waste treatment charge which is calculated by the city Electricity Works. This fee is paid by the waste collection agency to the Electricity Works as it is collected from households.

Certain forms of commercial and industrial waste are also treated if they are similar to household waste. The rates for this waste are the same as for household waste.

The collection and treatment charges are calculated for a three-year period on the basis of recent costs and forecasts of the likely evolution of these costs.

The principle is that all costs are to be met by the producers of the waste and there is to be neither subsidy nor profit by the city.

As of January 1, 1971, private householders paid 84 DM ($24.13) a year to have a 110 liter (3.9 cu ft) box emptied once every week. This increased to 115.20 DM ($43.15) per year in October 1973, and to 134.40 DM ($53.33) per year in January 1976 (Table 5.4). The additional charges for larger containers are proportionally smaller, but distance is not a factor in the rates.

The Electricity Works calculates the real direct and indirect costs of burning waste over a three-year period and deducts the value of the heat produced by the burning of these wastes. The difference is credited to the Electricity Works and paid from the fees collected by the waste collection agency. The treatment fees are not recalculated at the same time as the changes in the collection charges, and usually begin one year after the collection charges have been established for a 2-3 year period.

Political Structure of Solid Waste Collection and Treatment

Figure 5.2 outlines the chain of command for administrative and operational control of the collection and treatment of solid waste in Munich. The city is run by its own government under the control of a mayor and several deputy mayors. They sit with other city counselors in a committee and supervise many administrative subdivisions called Referats. One of these is the Werksreferat. This organization groups city transportation, gasworks, waterworks, city baths, and finally the Electricity Works.

The Electricity Works is broken down into an electricity section and a heat section. The former generates and distributes electricity, the latter operates the thermal power stations and the central heating plants and also incinerates waste. Within this heat group, there are an operating manager and a planning manager. The first controls all incinerator plants. The second supervises the expansion plans for new plants.

TABLE 5.4: RECENT WASTE COLLECTION AND TREATMENT CHARGES IN MUNICH (in Deutsche Marks)

	Total Collection and Treatment Charges[1]			Treatment Charges Alone[2]			
Container Sizes	**1/1/71 to 9/30/73**	**10/1/73 to 12/31/75**	**1/1/76 to Present**	**Up to 12/31/69**	**1/1/70 to 12/31/72**	**1/1/73 to 12/31/75**	**1/1/76 to Present**
110 litre box per week	7/mo 84/yr	9.60/mo 115.20/yr	11.20/mo 134.40/yr	0.80 DM per cubic meter, which was the old cost of landfilling	10.80 DM per metric ton burned	18.80 DM per metric ton burned	Not yet calculated
1.1 cubic meter box per week	58.50/mo 702/yr	80.20/mo 962.40/yr	89.50/mo 1,074/yr				
11 cubic meter box per week	475/mo 5,700/yr	650/mo 7,800/yr	650/mo 7,800/yr				
22-25 cubic meter box per week	—	—	1,105/mo 13,260/yr				
8 cubic meter box of compressed waste per week	605/mo 7,260/yr	780/mo 9,360/yr	780/mo 9,360/yr				
Smaller size pressed container waste per m^3 per week	75/mo 910/yr	45/mo 540/yr	45/mo 540/yr				

1 Charges actually paid by households and businesses to the city collection agency.
2 Credited to the electricity works by the city for waste treated by the electricity company.

Conversion rates:

Year	DM/$
69	3.69
70	3.64
71	3.48
72	3.19
73	2.67
74	2.59
75	2.46
76	2.52

FIGURE 5.2: ADMINISTRATIVE STRUCTURE FOR SOLID WASTE DISPOSAL IN MUNICH

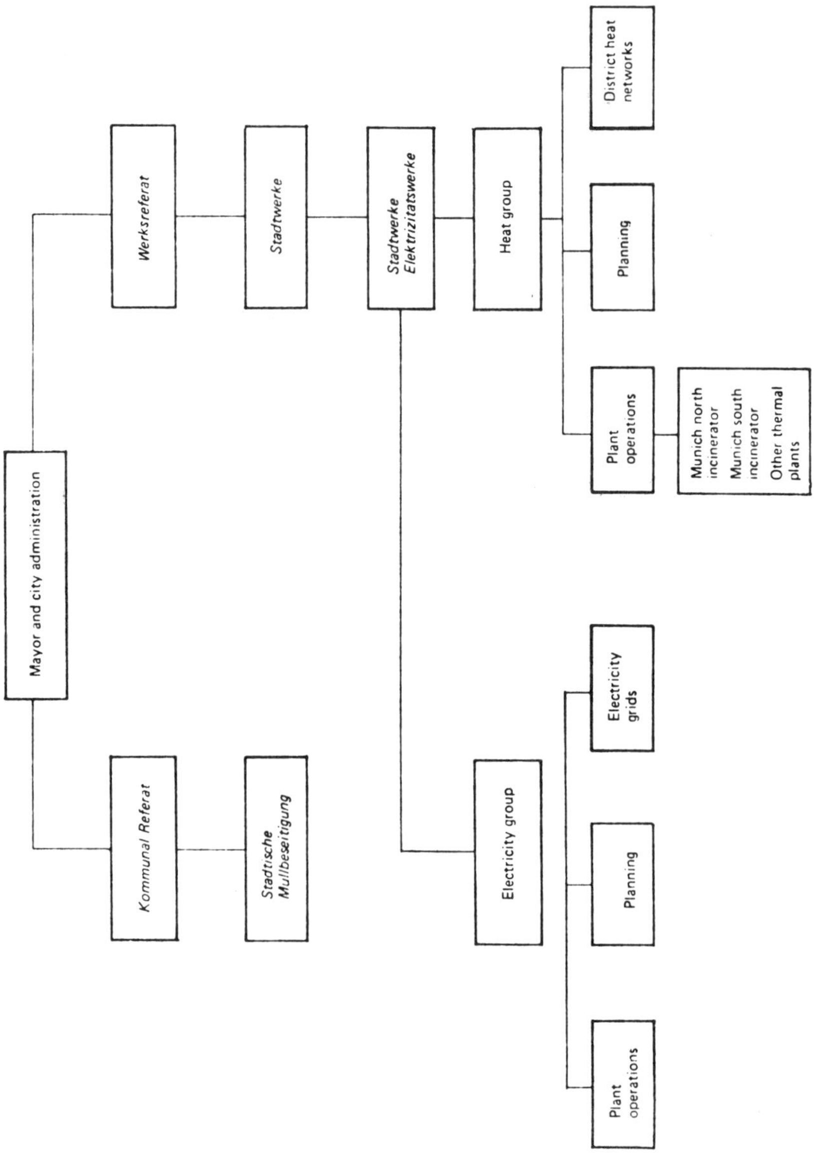

THE DEVELOPMENT OF THE SYSTEM

Prior to the decision to burn municipal waste, the city of Munich dumped its waste in a large open area to the north of the city center. During and after World War II, a private recycling company used to hand pick through this waste and recover various types of materials for resale.

In the early fifties, this landfill was closed for economic reasons and all waste was dumped in an area which eventually became known as the "garbage mountain" for its prodigious size. This mountain became more and more offensive, often catching fire, and generally the source of particulate matter that would be picked up by the wind and spread throughout the vicinity. Eventually there was insufficient dirt to cover the mountain if future deposits were made at the same rate. Thus, incineration seemed imperative.

At the same time, the city was engaged in planning a new thermal electric power station for the north of the city which would also provide district heating for a new development area. Having to solve two different problems at the same time led the city's political leaders to think that there could be some common solution which would not only get rid of the waste (at least in part) but also provide some fuel value. The political leaders and the head of the Werksreferat decided to have the Electricity Works incinerate waste along with fossil fuels in a combined plant for electricity generation and district heating.

No convenient land was available for this new facility within the city limits, but land was available just outside along the Isar Canal. This location, to the east of the city, was especially suitable because the prevailing wind blows from the west. The canal would provide the necessary water and there was room for a large fleet of trucks to bring waste.

The head of the Munich Werksreferat, having decided upon this solution, took a team to visit municipal incinerator plants throughout Europe. He was particularly concerned with the problems of smell and particulate matter in the air.

Siemens, a Munich-based company (and the eventual supplier of the electrical equipment), assisted the city in the planning for the new units. It was decided to order units designed for the efficient generation of electricity with the waste burning capacity as a secondary consideration. Bensen boilers were specified. It was also decided to order two incinerators connected to a single turbo-generator so that either furnace could be shut down for cleaning and maintenance while keeping the electrical generation in process. A request for proposals was issued and responses were received from a number of suppliers including Von Roll, Babcock, Volund, and KSG. Martin, another Munich-based firm, was an obvious consideration for combustion equipment.

The final decision-making process is unknown, but eventually the city itself contracted with individual suppliers for the first unit of the Munich North I plant. Siemens received the general contract for construction, etc. Martin was selected to supply the grate, Babcock the boiler, and Lurgi the electrostatic precipitators. Siemens, the general contractor, also supplied the turbo-generators. This plant was ordered in 1960 and the first unit, IB, was commissioned in June 1964. For the second unit, IA, the city decided to have the individual component suppliers work as subcontractors. The grates were selected and delivery made through the boiler suppliers. Martin thus became a subcontractor to Babcock for IA.

No alternatives for the operation of the plant were considered, as it was decided at the beginning that it would be integrated into and controlled by the Electricity Works. The city arranged the financing, partly from its own resources, and partly from bank loans.

The decision was initiated by the planning section of the Electricity Works' heat group. It was then approved all the way up to the City Council which gave the final approval. The city architects also approved the plans. A public announcement of the planned investments was made in the press, and the public had the right to comment on these projects. There was no public referendum or organized participation in the decision.

Even while the first two units of the Munich North plant (IA & IB) were being built, it was obvious that the capacity would not match the expected rise in the volume of waste collected. Also, as new areas were attached to the district heating networks, the amount of waste rose both in volume and in heat content (because paper and other burnable matter was no longer burned in homes).

By 1963, when planning for the expansion of the plant began, newer and larger turbines were available. It was decided to install a 120 MW turbine which meant that a larger boiler was required. The unit chosen was based upon an entirely new concept: burning waste and fossil fuel together in the same furnace with a lower relative contribution of heat from the waste, and with a much larger grate than had ever been used before.

Early experience of the North I units indicated that there were no major cleaning problems as a result of the burned waste, and that higher temperatures were possible from the burning waste. As in the case of the original North units, there was a conscious decision on the part of the city Electricity Works and municipal authorities to continue experimenting with radically new plant designs.

Bids were solicited and again there were several responses. Eventually Babcock again won the boiler contract and Martin was selected for the grates. This time, however, AEG, rather than Siemens, was selected for overall control of the project and for the supply of the turbines and generators. One reason was pressure from the federal government to provide work for AEG, as this company was based in Berlin and it was politically important for that city to thrive.

North II would be larger than North I, but would occupy the same building. This new unit was commissioned in May 1966.

At about the time the new North unit (North II) was commissioned, it was again necessary to plan an expansion of Munich's waste burning capacity. At the same time, the city's major priorities were to provide additional capacity for electricity generation and a district heating plant for a new region in the southeastern portion of the city at Perlach. The extra waste treatment capacity was needed because the actual operational capacity of the early North plants was lower than originally hoped. This was due both to higher heat content in the waste and to corrosion problems which were discovered. There was also increasing pressure to have the city accept more commercial and industrial refuse.

Once again an entirely new design was chosen. Two large incinerators were planned which would have completely separate waste and fossil fuel furnaces. The waste furnace served as an economizer unit for the fossil fuel furnace, but the latter was designed to operate totally independently if the waste furnace had to be shut down for any reason.

The North II plant was just coming into operation and there was a fear that the corrosion problems which were now quite evident and serious at the North IA and IB units would also occur on the North II plant. Thus, the North II design (which eventually was found to work quite well) was rejected for the South plants. Siemens was again selected to be the general contractor and the supplier of the turbo-generator unit. Martin was again selected to provide the grates and Babcock the boilers.

Munich South V was commissioned in April 1969 and South IV in July 1971. There are some slight design differences between these two units, mainly related to the size of the economizer in the waste-fired boiler and the size of the main boiler.

TECHNICAL INFORMATION: MUNICH NORTH I (A and B)

The Basic System

The Munich North I incinerator complex consists of two Babcock/Martin furnaces located side by side, fed by common overhead cranes. Both units feed a common turbo-generator and both discharge gases into a common stack. This system cogenerates heat and electrical energy.

Figure 5.3 shows a cross section of one of the furnace units. These Benson boilers, equipped with Martin reverse acting grates, have a nominal capacity

FIGURE 5.3: SECTION DRAWING OF MUNICH NORTH I INCINERATORS

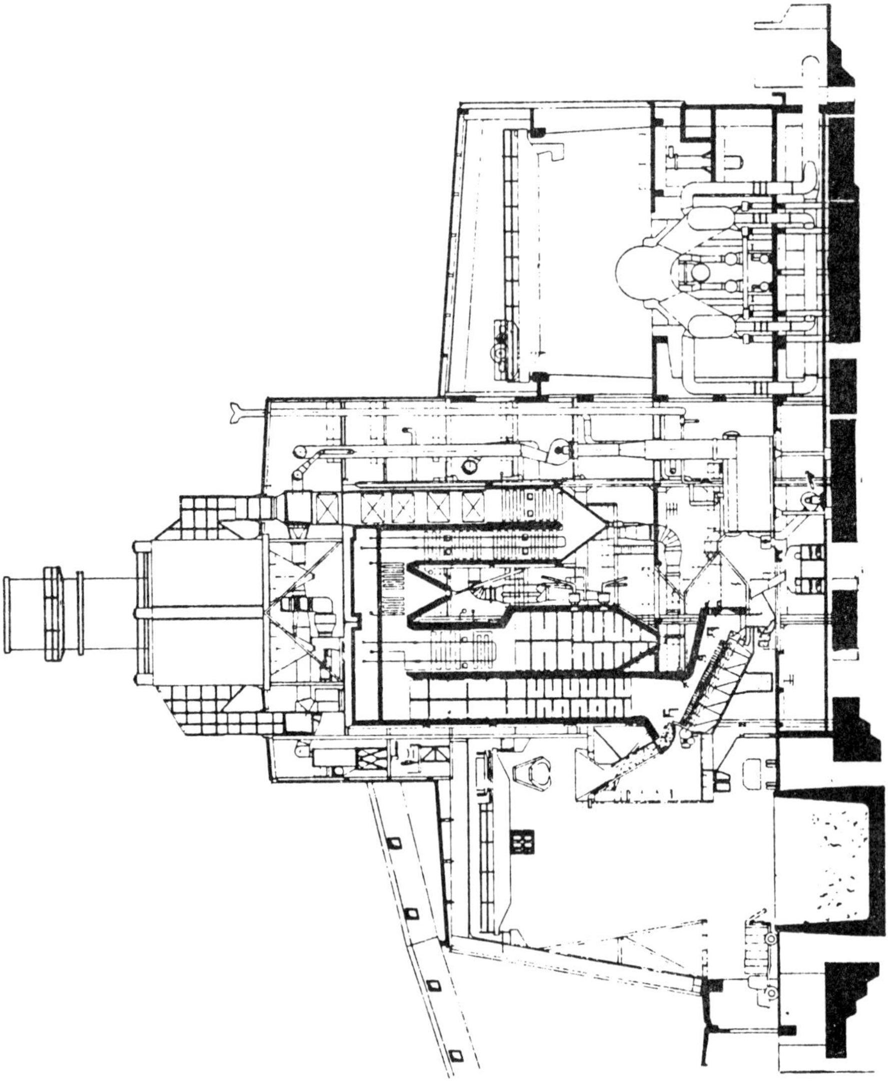

of 25 metric tons (27.5 short tons) per hour for waste with heat value from 900-1,800 kcal/kg (1,619-3,238 Btu/lb). Steam at 184 bars (181.5 atm) and 540°C (1000°F) feeds a 68 MW turbine. A reheat phase is incorporated between turbine stages, and heat is also supplied to heat exchangers for producing 120°C (248°F) steam for district heating.

Waste is fired in the portion of the furnace which is on the left in this diagram, while pulverized coal is fired separately in the parallel chamber to the right of the waste firebox. The gases combine above the coal furnace and pass jointly through the two vertical portions of the boiler convection area. The gases finally pass through a Lurgi electrostatic precipitator before escaping through the stack.

The collection trucks deposit waste in the large receiving bin. Overhead cranes, operated from a room where the operator can see the load level in each furnace intake hopper, lift the waste to the loading chutes. There is an automatic feed mechanism at the bottom of the loading chute which injects a controlled amount of waste to be burned.

This waste falls upon the uppermost layer of the Martin grate and starts to burn. The reverse action of the grate elements insures a very good burnout by continuously agitating the waste as it falls under gravitational force toward the ash roller. This roller controls the rate at which the ash falls into the quencher unit. The ash quenchers, also designed by Martin, consist of troughs where the ash and water combine. The ash is then pushed out of this trough onto a rubber conveyor belt which brings it to a separate building where ferrous material is magnetically separated. The remaining ash is stored and then hauled by truck to the city dump where it serves as cover for other waste.

The Martin grate has six inlet areas under each grate run which control the amount of air. There is a certain amount of back pressure resisting this inlet air for each stage of the grate to insure that uniform burnout takes place across the width of the grate. An air blowdown in the inlet manifold periodically removes the accumulated ash.

The plant site is 12 meters (39 ft) higher than the Isar Canal from which the condenser cooling water is drawn. A pump house contains three 12,000 m^3/hr (3,170,400 gal/hr) pumps, two of which are on line to service the entire installation (the two units of North I and the larger North II plant). The third pump is a reserve. The generator has a maximum output of 68 MW.

The plant was designed to operate in several different modes. The boiler can be operated at 100% capacity burning only coal. While there are evaporator and reheater tubes in the waste-fired furnace, there are also evaporator tubes and reheater tubes in the coal furnace and these are sufficient to deliver the necessary heat. The boiler can operate at the designed ratio of 40% of the heat input from waste and 60% from coal.

Finally, the plant can operate with only the waste-fired portion. In this mode no electricity can be generated but heat can be delivered to the district heating plant.

In general, the boilers operate more than the waste-fired furnaces. For example, as of February 1, 1968, the IA boiler had operated 17,700 hours while the waste-fired furnace had operated only 13,500 hours. The ratio was similar for the IB unit.

Figure 5.4 shows the arrangement of the boiler elements in the North I units and traces the flow of feedwater through to final superheated steam and the reheat passage. Figure 5.5 shows the basic steam flow and feedwater flow phases of the North I boiler and turbine combination. The turbine unit is composed of three sections. Steam from the outlet of the first section is mostly reheated before going to the remaining two sections of the turbine. Steam is drawn off at six intermediary stages for various purposes in addition to the reheat stage from the first part of the turbine series.

Level of Automation

The Munich North Plant I is highly automated. The crane operator sits in an insulated and isolated cabin from which he controls the operation of the crane. He can see the level in the loading hoppers directly. The loading hoppers automatically feed waste to the furnaces on a mechanically controlled cycle. The rate can be varied by the furnace control operators in the central control room.

The stoker action is mechanically operated through gear boxes and arms powered by electric motors. These will eventually be replaced by hydraulic units. The stoker speed is also controlled automatically from the control room.

Oil mist lubrication of mechanical parts is controlled from a central unit and automatically applied where needed regularly. The ash roller speed is adjusted for proper operation of the ash quenching process and/or control of delay time on the grate.

The air flow rate under the six areas of the grate runs is adjusted twice a year for winter and summer operation. It then runs by itself but is monitored from the central control room and from a gauge under each furnace.

Incineration requires very little direct labor on the furnace. One man is needed to watch for bulky items coming out of the ash hoppers to keep from fouling these units. Otherwise, apart from the men in the crane room and in the ash treatment center, almost all of the activity takes place in a central, air-conditioned room.

System Design Redundancy

There is quite a bit of redundancy in the North I plant. In the largest sense, the

FIGURE 5.4: BOILER TUBE ARRANGEMENTS IN THE MUNICH NORTH I UNITS

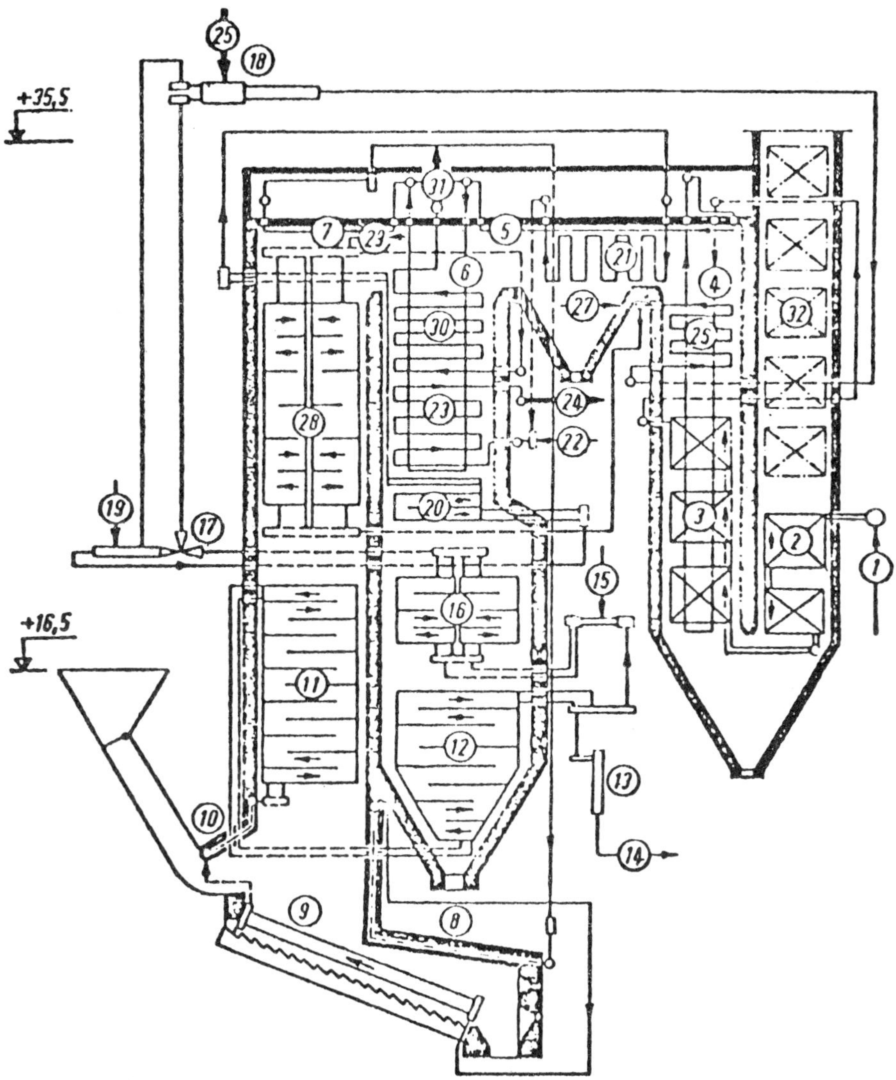

1. Feedwater entry
2. Economizer Unit 3
3. Economizer Unit 2
4. Support tubes 2
5. Ceiling wall tubes
6. Support tubes in coal furnace
7. Ceiling wall tubes
8. Grate ceiling tubes
9. Grate wall tubes
10. Grate entry tubes
11. Evaporator tubes
12. Evaporator tubes
13. Start-up drum
14. Auxiliary outlet
15. Emergency cooler
16. Superheater 1
17. Bypass ventilator
18. Auxiliary steam path
19. Cooler 2
20. Superheater 2a
21. Superheater 2b
22. Cooler 3
23. Superheater 3
24. Live steam outlet
25. Reheater entry
26. Reheater pass 1
27. Reheater cooler 1
28. Reheater pass 2
29. Reheater cooler 2
30. Reheater pass 3
31. Reheater outlet
32. Air preheater tubes

FIGURE 5.5: HEAT FLOW IN THE MUNICH NORTH I UNITS

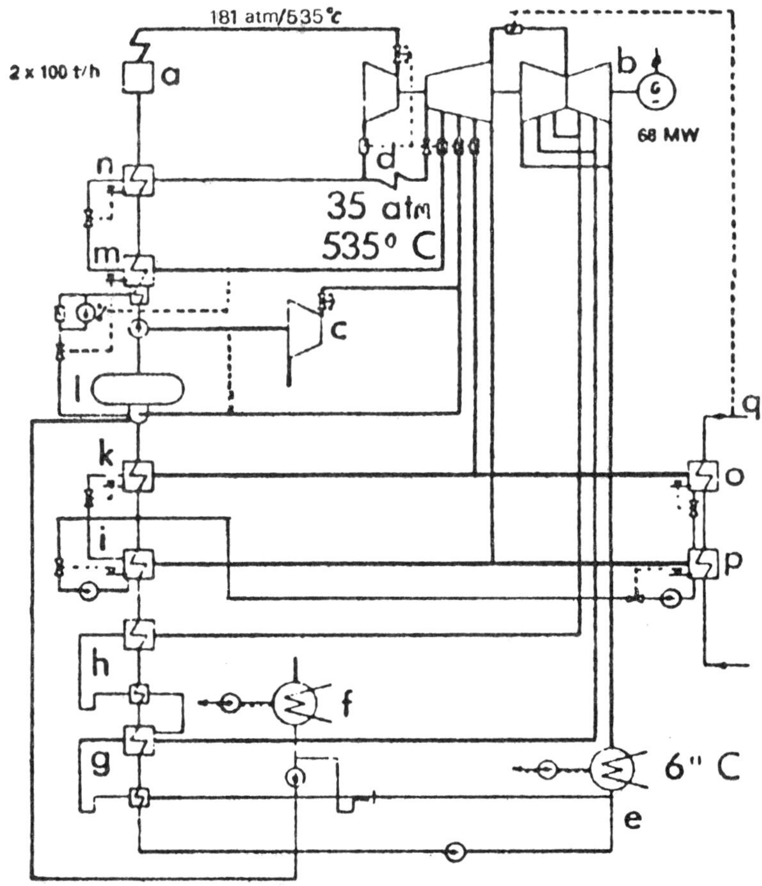

a. Boiler	i. Low pressure feedwater heater
b. Turbo-generator	k. Low pressure feedwater heater
c. Turbine for feedwater pump	l. Feedwater tank and deaerator
d. Reheating in boiler	m. High pressure heater II
e. Main condenser	n. High pressure heater I
f. Feedwater pump turbine condenser	o. District heating heat exchanger
g. Vacuum water heater	p. District heating heat exchanger
h. Low pressure feedwater heater	q. District heating hot water

presence of two incinerators feeding one turbine unit is a major redundancy for electricity generation. There are three cranes to insure that crane failure will not halt operation (actually, counting those of the North II unit in the same building, there are five cranes).

West German boiler codes are strict in requiring redundancy for pumps and accessories. The feedwater pumps must be at least 100% redundant throughout. For the main feedwater, the principal turbine-driven full load pump is supplemented by two half load pumps powered by three-phase electricity.

System Performance History

Table 5.5 shows the operating characteristics of the two units of Munich North I. From the beginning of operations to the end of 1975, a total of 1,239,043 metric tons (1,304,885 short tons) of waste were burned. The average heat value of the waste increased from an initial 1,200 kcal/kg (2,160 Btu/lb) to over 1,800 kcal/kg (3,240 Btu/lb).

The IA unit has accumulated 38,030 hours of service while the IB unit (the first to enter in service) has accumulated 40,920 hours, as of the end of 1975. Neither unit, however, has ever been used for more than 52% of the number of hours theoretically available each year. This is due to a variety of reasons which will be discussed later.

The rate of incineration has varied considerably and declined steadily from a maximum of an average of 20 metric tons (22 short tons) of waste per hour of service in 1965 to only about 14 metric tons (15.4 short tons) per hour in 1975. This is also due to a variety of reasons, including the higher heat value of the waste and the corrosion which occurs at high rates of heat release.

Materials and Energy Balance

The Electrical Works personnel watch the materials and energy balance closely at all of their installations. Daily, monthly, and yearly energy balances are calculated. Material balances are calculated in terms of energy contributed each day. Ash content and scrap metal recovery calculations are made less frequently.

For the daily energy balance (see the daily operating report, Table 5.10), the calculation is based upon known heat values of inputs of fossil fuels and from the known amount of heat recovered from the waste. An assumed heat content is used to estimate the amount of waste burned during the day. The annual balance takes the weight of the waste from known delivery volumes and calculates the actual heat content using assumed values for furnace efficiencies and for some of the losses.

Table 5.6 shows the most recent annual energy balance for North I (units IA and IB combined) for 1975. The average heat input from waste was 25.6% of the total heat intake of the plant.

TABLE 5.5: ANNUAL OPERATING CHARACTERISTICS OF MUNICH NORTH INCINERATORS IA AND IB

Year	Waste Burned (metric tons)*			Average Heat Value (kcal/kg)**		Hours in Service				Average Metric Tons Per Hour	
	IA	IB	Total	IA	IB	IA	%	IB	%	IA	IB
1964	19,401	29,839	49,240	1,178	1,178	1,057	12	1,624	19	18.4	18.4
1965	86,513	91,498	178,011	1,116	1,116	4,336	49	4,579	52	20.0	20.0
1966	81,133	74,892	156,025	1,394	1,394	4,355	50	4,020	46	18.6	18.6
1967	59,471	55,560	115,031	1,515	1,515	3,615	41	3,374	39	16.5	16.5
1968	54,034	68,771	112,805	1,478	1,478	3,137	36	3,985	45	17.2	17.3
1969	39,918	47,621	87,539	1,612	1,612	2,946	34	3,513	40	13.6	13.6
1970	51,364	54,979	106,343	1,594	1,594	3,446	39	3,690	42	14.9	14.9
1971	30,039	28,064	58,103	1,695	1,695	2,269	26	2,120	24	13.2	13.2
1972	47,211	50,736	97,947	1,782	1,782	3,745	43	4,032	46	12.6	12.6
1973	55,855	62,647	118,502	1,793	1,793	3,988	46	4,473	51	14.0	14.0
1974	35,967	47,066	83,033	1,712	1,712	2,534	29	3,316	38	14.2	14.2
1975	36,090	30,374	66,464	1,836	1,836	2,602	30	2,194	25	13.9	13.9
Total	**596,996**	**642,047**	**1,239,043**			**38,030**		**40,920**		**15.7**	**15.7**

* To convert metric tons to short tons, multiply by 1.1.
**To convert kcal/kg to Btu/lb, multiply by 1.8.

TABLE 5.6: ENERGY BALANCE FOR THE MUNICH NORTH I INCINERATOR IN 1975

Inputs	Metric System	English Equivalent
Coal		
Amount consumed	6956.5 metric tons	7663 short tons
Heat content	6905 kcal/kg	12,430 Btu/lb
Total heat	48,033 Gcal	1.91×10^{11} Btu
% total heat in	10	
Natural Gas		
Amount (volume)	34,369 thousand m^3	1213 million cf
Heat value	8878 kcal/m^3	997.8 Btu/cu ft
Total heat	305,132 Gcal	1.21×10^{12} Btu
% total heat in	64	
Oil (light oil)		
Amount consumed	183 metric tons	202 short tons
Heat value	10,200 kcal/kg	18,361 Btu/lb
Total heat	1867 Gcal	7.4×10^9 Btu
% total heat in	0.4	
Waste		
Amount consumed	66,464 metric tons	73,214 short tons
Heat value	1836 kcal/kg	3354 Btu/lb
Total heat	122,040 Gcal	4.84×10^{11} Btu
% total heat in	25.6	
Total heat input	**477,072 Gcal***	**1.89×10^{12} Btu***

Outputs	Metric System	English Equivalent
Electrical energy		
Total produced	173,499 MWh	
Delivered energy	160,148 MWh	
Self-consumed energy	13,351 MWh	
Self-consumed heat	15,602 Gcal	6.19×10^{10} Btu
Total heat in electricity and self usage	153,304 Gcal	6.08×10^{11} Btu
Total heat/heat in	32.1%	
District heat energy out (steam)	38,120 Gcal	1.51×10^{11} Btu
% total heat in	8	
Boiler losses (% heat in)	12.7	
Startup/shutdown losses (% heat in)	2.9	
Condenser and other losses (%)	44.2	
Total losses (% heat in)	59.8	
Total losses	285,648 Gcal	1.13×10^{12} Btu
Specific heat required for electricity production		
Fossil fuel only	2646 kcal/kWh	10,500 Btu/kWh
Waste only	3433 kcal/kWh	13,624 Btu/kWh

*Including 13,003 Gcal (5.16×10^{10} Btu) in a small separate part of the district heating circuit.

During the year, it took an average of 2,675 kcal (10,615 Btu) of heat to produce one kWh of electricity. If only fossil fuels had been used it is estimated that 2,645 kcal (10,500 Btu) would have been required while if only waste had been used 3,433 kcal (13,023 Btu) would have been required.

Using a conversion factor of 860 kcal/kWh, the actual operating efficiency of the plant with combined waste and fossil fuel firing was calculated at 32.2%. Firing only waste, the efficiency would have been 25.1% while firing only fossil fuel the efficiency would have been 32.5%.

The acceptance test values, measured boiler efficiency under various firing conditions, and an initial energy balance for a clean boiler are shown in Table 5.7. The anticipated efficiency for waste only was 27.7%; actual performance in 1975 was 25.1%.

TABLE 5.7: MEASURED BOILER EFFICIENCY AND ENERGY BALANCE OF MUNICH NORTH INCINERATORS IA AND IB

	Fuel Supplied				
	100% Coal	60% Coal	40% Waste		Mixture: 60% Coal 40% Waste
Boiler Losses and Boiler Efficiency* (percent)					
Unburned matter	2.0	3.3	6.5	11.5	6.6
Heat in ash	–	–	5.0		
Flue gas heat	5.6	4.7	15.0		8.8
Radiation heat	0.4	0.6	1.1		0.7
Total boiler losses	**8.0**	**8.6**	**27.6**		**16.1**
Boiler efficiency	**92.0**	**91.4**	**72.4**		**83.9**
Energy Balance for a Clean Boiler in Full Operation					
Turbine efficiency (percent)	45.0	(45.0	45.0)		45.0
Self-consumption removed (percent)	92.5	(91.0	93.0)		91.8
Boiler efficiency (percent)	92.0	(91.0	72.0)		83.0
Combined (percent)	**38.2**	**(37.3**	**30.2)**		**34.2**
Heat-in per kWh (kcal)**	2250	(2300	2850)		2520
Heat-in including an 8.5% factor for unclean tubes, part loads, and startup/shutdown (kcal)	2450	(2500	3100)		2740
Final expected efficiency (percent)	**35.1**	**(34.4**	**27.7)**		**31.4**

* Measured during acceptance tests for the Munich North I plant.
**One kcal = 3.968 Btu.

Maintenance History, Repairs and Modifications

The North I units were found to be severely corroded after about a year of operation. The most prevalent theory was that this was caused by HCl gas released in burning the waste. However, there were also problems from the alternating oxidizing/reducing atmosphere in the firebox and from the variations in acidic/basic composition of the gases in different parts of the air above the grates.

A partial solution, particularly for the visible corrosion on the lower part of the waterwall tubes in the waste-fired furnace, was to fix studs to the outer tube walls and hang silicon carbon insulation on them.

Corrosion was also found on the reheater tubes in the waste-fired furnace but nothing could be done to protect these. Additional corrosion was found in the superheater tube section and in the downdraft convection channel. This downdraft corrosion was attributed to overly high flue gas flow and particulate matter eroding and exposing the tube surfaces. The particulate matter from the coal fired boiler was too great because this furnace seems to be too small. Too much carbon ash is produced.

As a result, it has been necessary to replace tubes throughout the boiler far more often than was intended. Some of these have been thickened to prolong their lives, at a cost of some decrease in their heat absorption capacity. It has been found necessary to reduce the rates of steam generation and waste incineration to reduce the danger of too much corrosion.

Nowadays, since the North II unit performs much better than the North I (A and B), North I is shut down often, sometimes every evening and usually each weekend. This allows frequent inspection of the tubes and permits some minor repairs.

Another design problem is that Babcock did not install waterwall tubes right down to the level of the grate. Thus, there are refractory walls part of the way up inside each waste firebox. This causes much more intense heat and subsequent ash fusing on the refractory walls, as well as higher gas temperatures at the evaporator tubes.

In addition to the changes made to protect the tubes in the waste firebox, the two North I incinerators have been modified in other ways. Gas burners were added in 1974 to allow the furnaces to burn cheaper fuel (gas from the Soviet Union is cheaper than West German coal). The system has also been modified to permit the bypassing of tubes in the waste portion, if necessary, for a short period so that waste burning and turbine operation can continue until the next planned shutdown. The fans were found to be too small and were replaced.

Tables 5.8 and 5.9 show the availability of the waste grates and furnaces for the IA and IB incinerators in 1975. Planned repair and maintenance accounted for 28 and 25% of the total available time respectively. Unplanned stoppages accounted for only 1 and 2%. The furnaces were actually in operation for 30 and 25% of the available hours, leaving them in reserve for 41 and 48% of the time (hours when they could have been used but weren't). The furnaces were left idle either because of lack of demand for electricity, or because of lack of waste to burn.

TABLE 5.8: AVAILABILITY OF MUNICH NORTH INCINERATOR IB IN 1975 BY MONTH, (hours)

Month	Hours in Service	Planned Stoppages: Repair and Maintenance	Unplanned Stoppages	Reserve Hours	Total
January	–	–	–	744	**744**
February	–	–	–	672	**672**
March	–	744	–	–	**744**
April	–	720	–	–	**720**
May	–	720	–	24	**744**
June	558	–	–	162	**720**
July	253	–	72	420	**744**
August	281	–	7	455	**744**
September	113	–	–	607	**720**
October	276	–	–	468	**744**
November	450	–	–	269	**720**
December	263	–	73	408	**744**
Total	**2194**	**2184**	**152**	**4229**	**8760**
Percent	**25**	**25**	**2**	**48**	**100**

TABLE 5.9: AVAILABILITY OF MUNICH NORTH INCINERATOR IA IN 1975 BY MONTH, (hours)

Month	Hours in Service	Planned Stoppages: Repair and Maintenance	Unplanned Stoppages	Reserve Hours	Total
January	235	–	–	509	**744**
February	230	–	–	442	**672**
March	–	744	–	–	**744**
April	–	720	–	–	**720**
May	–	744	–	–	**744**
June	106	264	–	350	**720**
July	464	–	–	280	**744**
August	326	–	24	394	**744**
September	446	–	48	226	**720**
October	304	–	–	440	**744**
November	183	–	–	537	**720**
December	308	–	–	436	**744**
Total	**2602**	**2472**	72	**3614**	**8760**
Percent	**30**	**28**	**1**	**41**	**100**

Both boilers in IA and IB actually operated longer than the waste grates (being fired without waste at times), but both had much higher unplanned stoppages. Unplanned stoppages on boiler IA accounted for 11% of total hours per year. The equivalent figure for boiler IB was 17%. As Table 5.5 shows, units IA and IB have never been used more than 52% of total yearly hours in any year since they were built.

Planned maintenance is determined by several institutional requirements. All West German boilers must undergo a major inspection every two years as part of the West German boiler code and certification program. The insurance companies require a major inspection of the turbines every three years. In addition, an annual 4-5 week period is scheduled for repairs to the grate and for any necessary modifications of the tubes, additional insulation, etc.

The heavy corrosion in the waste portion of the furnace means that major repair programs have to be planned for replacement of refractory brick, secondary air nozzles, etc. Since these units are shut down nearly every weekend, there is frequent opportunity to perform smaller repairs without having to shut down during planned operational periods.

Table 5.10 is a daily operation report for the two units on July 27, 1976. The IA unit burned coal, some natural gas, and waste for the full 24 hour period. The IB unit burned only coal for 14.5 hours.

The actual operating rates and values are lower than the design rates shown in Table 5.1. The average steam rate for IA was 91.5 metric tons (100.8 short tons) per hour instead of 100 metric tons (110 short tons) per hour. Feedwater temperature was actually 232°C (450°F) instead of 250°C (482°F). The temperature of the flue gas was 174°C (345°F), rather than 140°C (284°F) which had been anticipated. Waste was burned at a rate of 14.8 metric tons (16.3 short tons) per hour instead of 25 metric tons (27.5 short tons) per hour.

The plants are rarely run at a constant rate for the 24 hour period because of fluctuations in the daily electrical load in the city of Munich. Thus even this daily operational report represents an average of output figures. Load requirements ran from a maximum output of 60 MW to an average rate for the day of 43 MW.

TECHNICAL INFORMATION: MUNICH NORTH II

The Basic System

The Munich North II incinerator is built around a high pressure and temperature Benson boiler fired by a combined waste and coal furnace. Steam from the boiler provides heat to a district heating system and also feeds a 112 MW turbine which generates electricity.

TABLE 5.10: DAILY OPERATING REPORT FOR MUNICH NORTH I INCINERATOR

	Unit	IA	IB	Total
Boiler/Incinerator				
Feedwater volume	metric tons	2,196.0	1,221.0	**3,407.0**
Feedwater temperature	°C	232.0	234.0	
Live steam volume	metric tons	2,196.0	1,221.0	**3,407.0**
Live steam pressure	bar	176.0	176.0	
Live steam temperature	°C	530.0	530.0	
Recovered heat/kg	kJ/kg	2,363.0	2,354.0	
Total heat in steam	GJ	5,189.2	2,850.6	**8,039.8**
Reheating				
–Volume	metric tons	2,019.0	1,130.5	**3,149.5**
–Pressure–in	bar	39.0	40.0	
–Temperature–in	°C	336.0	343.0	
–Cooling spray	metric tons	13.0	23.0	**36.0**
–Soot blowing	metric tons	16.0	5.0	**21.0**
–Pressure–out	bar	36.0	37.0	
–Temperature–out	°C	530.0	530.0	
Recovered heat (reheat) per kg	kJ/kg	466.6	476.0	
Heat in received steam	GJ	942.1	538.1	**1,480.2**
Total heat in steam	GJ	6,131.3	3,388.7	**9,520.0**
Flue gas temperature after air preheater	°C	174.0	165.0	
O_2 content after preheater	%	8.6	7.3	
Air temperature before aph	°C	42.0	46.0	
Boiler losses in				
–Flue gas	%	10.5	7.0	
–Ash heat	%	0.7	0.3	
–Unburned ash	%	–	–	
–Unburned flue ash	%	–	–	
–Radiation losses	%	1.3	1.4	
Boiler efficiency	%	87.6	91.2	**88.8**
Coal burned	metric tons	107.0	118.7	**225.7**
–Heat	GJ	3,225.5	2,577.6	**6,803.1**
Natural gas burned	KCBM I.N	33.5		**33.5**
–Heat	GJ	1,247.2		**1,247.2**
Waste burned	metric tons	356.1		**356.1**
–Heat (7,100 kJ/kg)	GJ	2,528.3		**2,528.3**
Total fuel heat	GJ	7,001.0	3,714.3	**10,715.3**
Hours in service	hrs	24.0	14.5	**38.5**
Turbine				
Steam received	metric tons	3,377.0*		
Live steam pressure	bar	176.0		
Live steam temperature	°C	530.0		
Specific steam consumed	kg/kWh	3.2		
Generator				
Electricity produced	MWh	1,040.0		
Electricity consumed internally	MWh	70.0		
Electricity delivered	MWh	970.0		
Maximum power output	MW	60.0		
Average power output	MW	43.3		
Specific heat consumed	kJ/kWh	11,046.7		

SOURCE: Munich Electricity Works.

NOTE: One metric ton = 1.1 short tons
One bar = 0.987 atmospheres
One GJ = 0.9486×10^6 Btu
One kJ = 0.9486 Btu
One kg = 2.2 pounds

To convert kJ/kg to Btu/lb multiply by 0.431.
(°C x 1.8) + 32 = F°

*Thirty tons to low pressure bank.

Figure 5.6 shows a cross section of the furnace and boiler. The furnace is equipped with Martin reverse acting grates (five runs in parallel) and has a nominal waste burning capacity of 40 metric tons (44 short tons) per hour. The boiler has a capacity of 365 metric tons (402 short tons) of steam per hour at 184 bars (181.5 atm) and 540°C (1000°F).

Waste is burned in the bottom portion of the furnace after being fed in through a loading hopper and controlled feeding chute. Pulverized coal is fired above the refuse grate from three separate firing stages, some or all of which are used, depending on the operating load of the plant.

This design is a radical departure from that of the first combined waste and coal units of the North I installation. Not only is the waste area substantially larger (five runs rather than three) and the waste firing volume 60% greater than in the earlier furnaces, but also the coal firing takes place in the same furnace. The new unit produces over three and a half times as much steam and the contribution of waste, although much larger in volume, has fallen relative to coal. At full output waste contributes only 20%, rather than 40% as in the earlier system.

The only boiler elements in the furnace area are wall tubes. The superheater and reheater, as well as the economizer bundles, are outside the firebox in the two convection passes.

The waste storage bin for the North II unit is an extension of the bin used for the two units of North I. A separate crane control room was built for this new furnace, but cranes from the North I unit can also be used, as the overhead travelling units are connected.

The hopper and feeding mechanisms for supplying waste to the grate are similar in design to the North I unit. The grate runs are exactly the same as in North I: 17 grate steps, eight of which are fixed and nine of which are moveable. The moveable steps are actuated mechanically by electrical motors and gear drives. The air inflow is also similar. There are six inlet zones beneath the grate where air, facing a back-pressure of 300 mm, is regulated in a distinct pattern to insure uniform burning across the five runs, despite the non-homogeneous nature of the waste. The air supply system is the same as in North I and also incorporates air blowdown to remove dust which falls in through the inlet dampers.

The waterwalls start eight meters (26 ft) above the grate. This, as will be explained, makes cleaning the furnace complicated.

North II shares the pumping station which lifts condenser cooling water from the Isar Canal for North I. When all three furnaces are in operation, two pumps of 12,000 m^3 (3,170,400 gal) per hour are on line with a third in reserve.

The generator has a basic rating of 140 MW with a cosine phi of 0.8 giving an effective output of a maximum of 112 MW.

FIGURE 5.6: SECTION DRAWING OF MUNICH NORTH II INCINERATOR

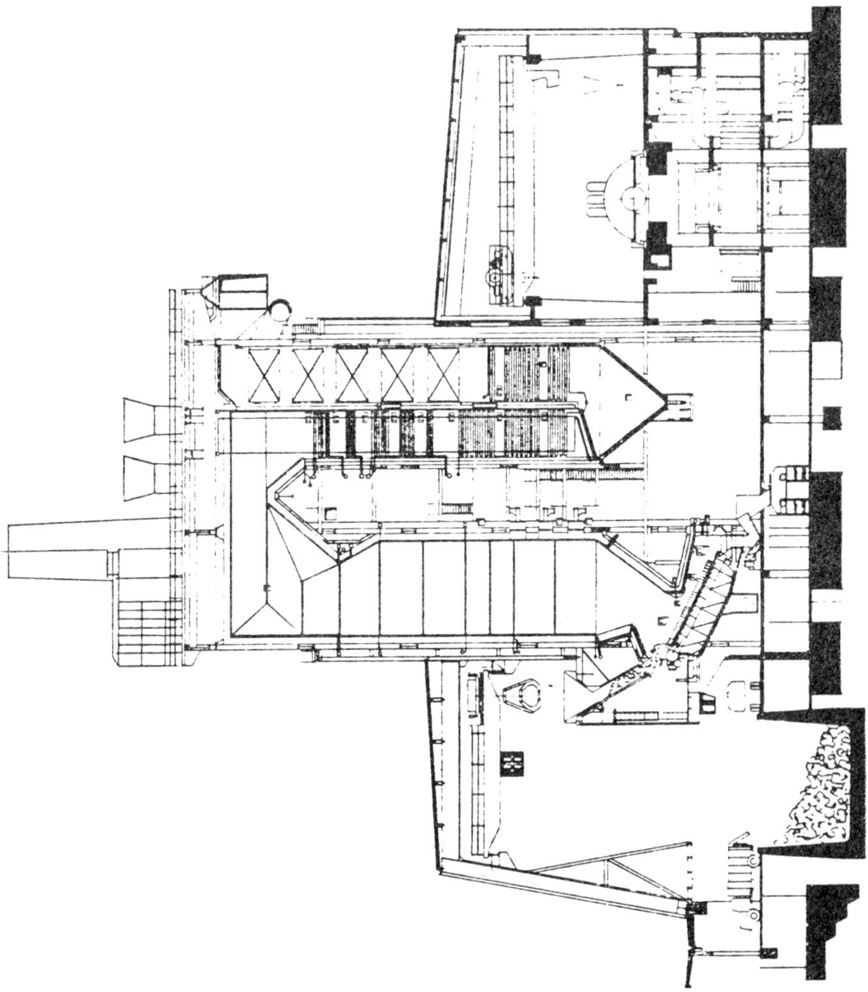

The plant can operate in several different modes at various load levels. Since waste and coal are fired in the same furnace area, the furnace can operate entirely on waste, or entirely on coal, or on a combination of the two. A maximum of 20% waste and 80% coal can be used at full rated output. It has been found that if the firing of this furnace is begun with coal and the waste firing is started afterwards, there is less corrosion.

The North II furnace is equipped with three Martin ash quenching units. The ash falls into a shallow pool of water, is quenched and pushed out an upward slope. The cooling water does not circulate but either evaporates up into the furnace, or is carried away as part of the ash (15% water after quenching).

Like those in North I, the boilers were designed and built by Babcock. The grate, the feeding hopper, and the ash quenchers are from Martin. The turbine is from AEG. The electrostatic precipitator is from Lurgi.

The Basic Steam and Feedwater Flow System

Figure 5.7 shows the basic steam and feedwater flow phases of the North II boiler and turbine combination. The turbine unit is composed of three sections. Steam from the outlet of the first section is mostly reheated before going to the remaining two sections. Steam is drawn off at six intermediary stages for various purposes in addition to the reheat stage from the first part of the turbine series.

Level of Automation

The North II incinerator is as highly automated as its predecessors. The remote control system is similar.

Ash quenching is also controlled automatically. A shovel cycle allows the ash unit to fill with water and a certain amount of ash to be quenched. There is then an expulsion cycle during which the quenched ash is pushed up a slope and then falls through a rough grate (which stops large items: logs, car doors, etc.) on to the ash conveyor belts. These belts continuously move the ash to the separate building where ferrous metal is electromagnetically removed and automatically loaded onto railway cars. The remaining ash is automatically conveyed to another building containing a bunker from which it is finally loaded onto trucks and hauled to the municipal dump.

As with North I, the West German boiler code requires backup capacity for all of the major feedwater pumps, condenser cooling pumps, etc. The cranes for North II can be shared with North I because the overhead rails are continuous. The North II furnace has its own stack and electrostatic precipitator.

There is, however, no crossover capability between the North I and II turbine units. Thus, if the North II furnace is down, the turbine cannot be fed by steam from the North I boilers. This lack of redundancy is explained by the extreme difficulty such a design would pose for the reheating phase of the steam flow.

FIGURE 5.7: HEAT FLOW IN THE MUNICH NORTH II UNITS

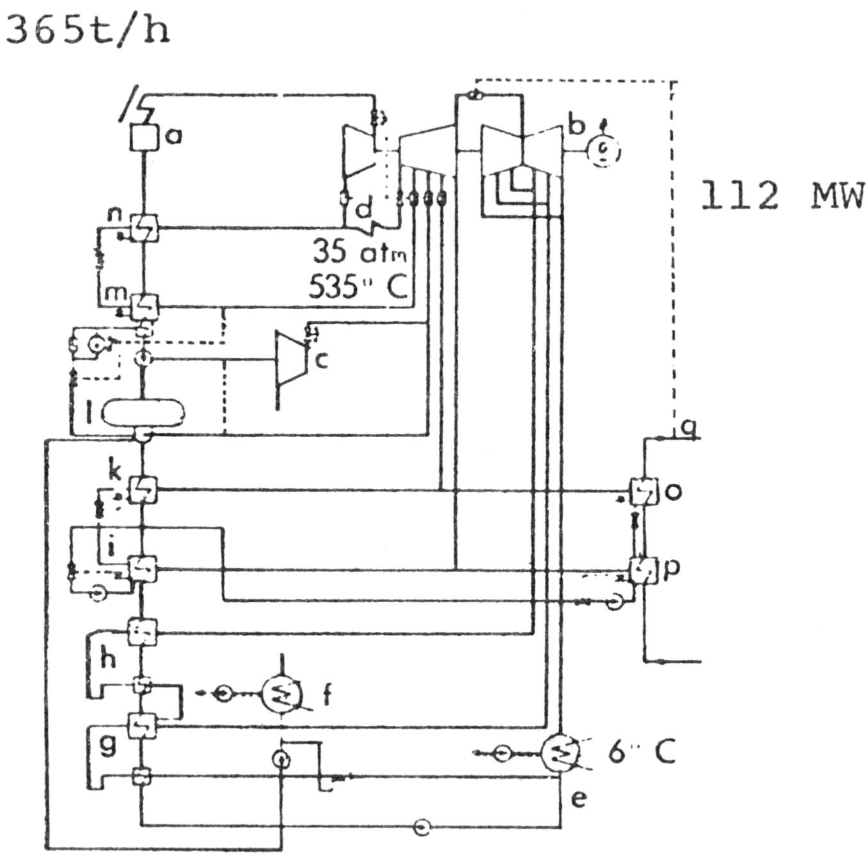

a. Boiler
b. Turbo-generator
c. Turbine for feedwater pump
d. Reheating in boiler
e. Main condenser
f. Feedwater pump turbine condenser
g. Vacuum water heater
h. Low pressure feedwater heater
i. Low pressure feedwater heater
k. Low pressure feedwater heater
l. Feedwater tank and deaerator
m. High pressure heater II
n. High pressure heater I
o. District heating heat exchanger
p. District heating heat exchanger
q. District heating hot water

System Performance History

Table 5.11 shows the annual operating characteristics of North II. From the beginning of operations through the end of 1975, a total of 1,367,081 metric tons (1,505,927 short tons) of waste were burned. The average heat value of the waste has increased from an initial 1,400 kcal/kg (2,519 Btu/lb) to over 1,800 kcal/kg (3,238 Btu/lb).

TABLE 5.11: ANNUAL OPERATING CHARACTERISTICS OF MUNICH NORTH INCINERATOR II

Year	Waste Burned (metric tons)*	Average Heat Value (kcal/kg)**	Hours in Service	Percent	Average Metric Tons Per Hour
1966	54,059	1,394	1,580	18	34.2
1967	159,978	1,515	5,040	58	31.7
1968	184,891	1,478	6,297	72	29.4
1969	156,519	1,609	6,186	71	25.3
1970	120,413	1,693	4,638	53	26.0
1971	154,113	1,693	6,404	73	24.1
1972	149,746	1,785	5,540	63	27.0
1973	106,635	1,709	4,721	54	22.6
1974	142,263	1,754	6,059	69	23.5
1975	138,464	1,875	5,928	68	23.4
Total	**1,367,081**		**52,393**		**26.1**

* To convert metric tons to short tons, multiply by 1.1.
**To convert kcal/kg to Btu/lb, multiply by 1.8.

By the end of 1975, the North II unit had accumulated a total of 52,393 hours in service. The maximum service that this unit has seen in any one year was in 1971 when the incinerator operated for 73% of the theoretical maximum number of operating hours.

The burning rate has varied considerably in terms of average number of tons of waste burned per hour of plant operation. From a high of 34.2 metric tons (37.6 short tons) per hour during the first year of operation, the average has dropped to about 23 metric tons (25.3 short tons) per hour during the last three years. This is only 60% of designed capacity.

Materials and Energy Balance Calculations

The daily energy balance for North II is calculated in the same manner as the daily energy balance for North I, see page 145.

Table 5.12 shows the most recent annual energy balance for the North II incinerator. The average heat input from waste in 1975 was 21.5% (vs a designed maximum of 20%).

TABLE 5.12: ENERGY BALANCE FOR THE MUNICH NORTH II INCINERATOR IN 1975

Inputs	Metric System	English Equivalent
Coal		
Amount consumed	134,495.2 metric tons	148,155.1 short tons
Heat content	6925 kcal/kg	12,466 Btu/lb
Total heat	931,400 Gcal	3.7 x 10^{12} Btu
% total heat in	77.2	
Natural gas	None	
Oil (light)		
Amount consumed	1469.1 metric tons	1618.3 short tons
Heat value	10,200 kcal/kg	18,361 Btu/lb
Total heat	14,985 Gcal	5.95 x 10^{10} Btu
% total heat in	1.3	
Waste		
Amount consumed	138,464 metric tons	152,527 short tons
Heat value	1875 kcal/kg	3375 Btu/lb
Total heat	259,659 Gcal	1.03 x 10^{12} Btu
% total heat in	21.5	
Total heat input	**1,206,004 Gcal**	**4.79 x 10^{12} Btu**

Outputs	Metric System	English Equivalent
Electrical energy out		
Total produced	489,760 MWh	
Total delivered	455,151 MWh	
Self-consumed energy	34,609 MWh	
Self-consumed heat	8960 Gcal	3.56 x 10^{10} Btu
Heat in electricity and self usage	430,079 Gcal	1.71 x 10^{12} Btu
Total heat as % of total heat in	35.7	
District heat energy out (steam)	132,360 Gcal	5.25 x 10^{11} Btu
% total heat in	11	
Boiler losses (% heat in)	11.6	
Startup/shutdown losses (% heat in)	0.4	
Condenser and other losses (% heat in)	41.3	
Total losses (% heat in)	53.3	
Total losses	643,605 Gcal	2.55 x 10^{12} Btu
Specific heat required for electricity production		
Fossil fuel only	2446 kcal/kWh	9707 Btu/kWh
Waste only	3015 kcal/kWh	11,965 Btu/kWh

On the average, the plant produced one kWh for a heat input of 2,463 kcal (9,774 Btu). Based on the efficiency differential for the burning of waste and of fossil fuel, only 2,446 kcal (9,707 Btu) would have been needed if only fossil fuels were used, while 3,015 kcal (11,965 Btu) would have been required if only waste were used.

Using a conversion factor of 860 kcal/kWh, the average plant conversion efficiency (heat to electricity) was 34.9%. The efficiency of fossil fuel conversion was 35.2%, while the efficiency of waste conversion was 28.5%.

Maintenance History, Repairs and Modifications

There has been much less corrosion and erosion in North II than in North I. The design of the North II firebox incorporated major changes in the conception of the burning area and air flow. The secondary air pattern was carefully designed to add more turbulence. This produced better burnout lower in the firebox and prevented the alternation of oxidizing and reducing conditions in the area of the wall tubing to a considerable extent. Corrosion has been observed in the lower part of the firebox near the secondary air inlet ports, but no major corrosion problems have been discovered in the superheater or reheater bundles, although some superheater tubes have now been replaced.

The most significant design problem has been with the waterwalls which do not extend down to the refuse grate. The lack of any heat absorption in the bottom part of the furnace means that more intense concentrations of heat occur, which leads to slag fusing on the refractory walls. Slag sometimes builds up to several feet in thickness. The furnaces eventually have to be shut down and cleaned out, a difficult and even dangerous process because loosening the lower layers of the fused soot can bring torrents of soot tumbling down from higher in the furnace.

The waterwalls in the furnace have recently been extended farther down and some more protected tubes have been added. This will lessen the problem of slag fouling the lower furnace walls. The tubes will be protected from the worst corrosion by silicon carbon coverings held in place by studs on the tubes.

Table 5.13 shows the availability of the Munich North II incinerator in 1975, by month. Planned repair and maintenance activities accounted for 19% of the total incinerator time. Unplanned stoppages amounted to only 1% of the time. For 12% of the time, North II was in reserve, either because there was not enough waste to burn or because there was no demand for electricity. The plant operated for 68% of the year. Combining the operation and reserve time, the plant was actually available 80% of the year.

Maintenance is planned around the same institutional requirements as are in effect for the North I units. Thus, external constraints set idle hours for the unit. These periods, naturally, are available for maintenance that may be otherwise necessary on other system components.

The grate usually requires three to four weeks worth of inspection and repair per year. However, stoppages on weekends or when the system is not required (reserve hours) provide opportunities to make inspections and carry out minor repairs which could otherwise later cause stoppages.

Table 5.14 shows the average operating characteristics of the North II plant on a typical day in 1976. On the 27th of July, the plant was in service for 24 hours and produced 5,370 metric tons (5,915 short tons) of steam. The average

TABLE 5.13: AVAILABILITY OF MUNICH NORTH INCINERATOR II IN 1975 BY MONTH (hours)

Month	Hours in Service	Planned Stoppages: Repair and Maintenance	Unplanned Stoppages	Reserve Hours	Total
January	620.5	–	11	112.5	**744**
February	507.0	–	18	147.0	**672**
March	660.5	–	–	83.5	**744**
April	661.5	–	–	58.5	**720**
May	617.0	–	–	127.0	**744**
June	–	696	–	24.0	**720**
July	–	744	–	–	**744**
August	565.5	–	–	178.5	**744**
September	612.0	–	74	33.5	**720**
October	480.0	168	–	96.0	**744**
November	625.0	–	2	93.0	**720**
December	578.5	72	–	93.5	**744**
Total	**5927.5**	**1680**	**105**	**1047.0**	**8760**
Percent	**68**	**19**	**1**	**12**	**100**

hourly rate was 224 metric tons (247 short tons), well under the theoretical maximum of 365 metric tons (402 short tons) per hour. The feedwater temperature was 243°C (469°F) instead of the designed temperature of 266°C (511°F). Live steam temperature was 530°C (988°F) instead of the designed temperature of 540°C (1004°F). Flue gas temperature was 195°C (383°F) and O_2 averaged 7.2%.

Waste was burned at a rate of 23 metric tons (25.3 short tons) per hour, and provided 23.6% of the total heat input (compared to the designed contribution of 20%). The electrical output during this 24 hour period was 1730 MWh, for an average of 74.5 MWh. The maximum power out during this period was 85 MW (compared to total capacity of 112 MW).

TECHNICAL INFORMATION: MUNICH SOUTH IV AND V

The Basic System

The Munich South incinerator complex consists of two Babcock/Martin furnace and boiler units sitting side by side, fed by common overhead cranes. Each furnace and boiler unit has its own turbo-generator, but they share a common stack.

Figure 5.8 shows a cross section of one of these furnace and boiler units. The South IV and V units are a radical departure from the designs of the furnaces

TABLE 5.14: DAILY OPERATING REPORT FOR MUNICH NORTH II INCINERATOR (July 27, 1976)

Boiler/Incinerator		
Feedwater volume	metric tons	5,370.0
Feedwater temperature	°C	243.0
Live steam volume	metric tons	5,370.0
Live steam pressure	bar	176.0
Live steam temperature	°C	530.0
Recovered heat per kg	kJ/kg	2,312.9
Heat in steam	GJ	12,420.2
Reheating		
–Volume	metric tons	4,874.6
–Pressure–in	bar	39.0
–Temperature–in	°C	336.0
–Cooling spray	metric tons	–
–Soot blowing	metric tons	36.0
–Pressure–out	bar	36.0
–Temperature–out	°C	530.0
Recovered heat (reheat) per kg	kJ/kg	449.8
Heat in reheater pass	GJ	2,192.5
Total heat in steam	GJ	14,612.7
Flue gas temperature after air preheater	°C	195.0
O_2 content after preheater	%	7.2
Air temperature before preheater	°C	50.0
Boiler losses in		
–Flue gas	%	9.6
–Ash heat	%	0.6
–Unburned ash	%	–
–Unburned flue ash	%	–
–Radiation losses	%	1.5
Boiler efficiency	%	88.3
Coal burned	metric tons	418.7
–Heat	GJ	12,621.7
Waste burned	metric tons	552.8
–Heat (7,100 kJ/kg)	GJ	3,925.0
Total fuel heat	GJ	16,546.7
Hours in service	hours	24.0
Turbine		
Steam received	metric tons	5,340.0*
Live steam pressure	bar	176.0
Live steam temperature	°C	530.0
Specific steam consumed	kg/kWh	3.0
Generator		
Electricity produced	MWh	1,730.0
Electricity consumed internally	MWh	100.0
Electricity delivered	MWh	1,630.0
Maximum power output	MW	85.0
Average power output	MW	74.0
Specific heat consumed	kJ/kWh	9,866.9

SOURCE: Munich Electricity Works

NOTE: One metric ton = 1.1 short tons
One bar = 0.987 atmospheres
One GJ = 0.9486×10^6 Btu
One kJ = 0.9486 Btu
One kg = 2.2 pounds

To convert kJ/kg to Btu/lb multiply by 0.431.
(°C x 1.8) + 32 = F°

*Thirty tons to low pressure bank.

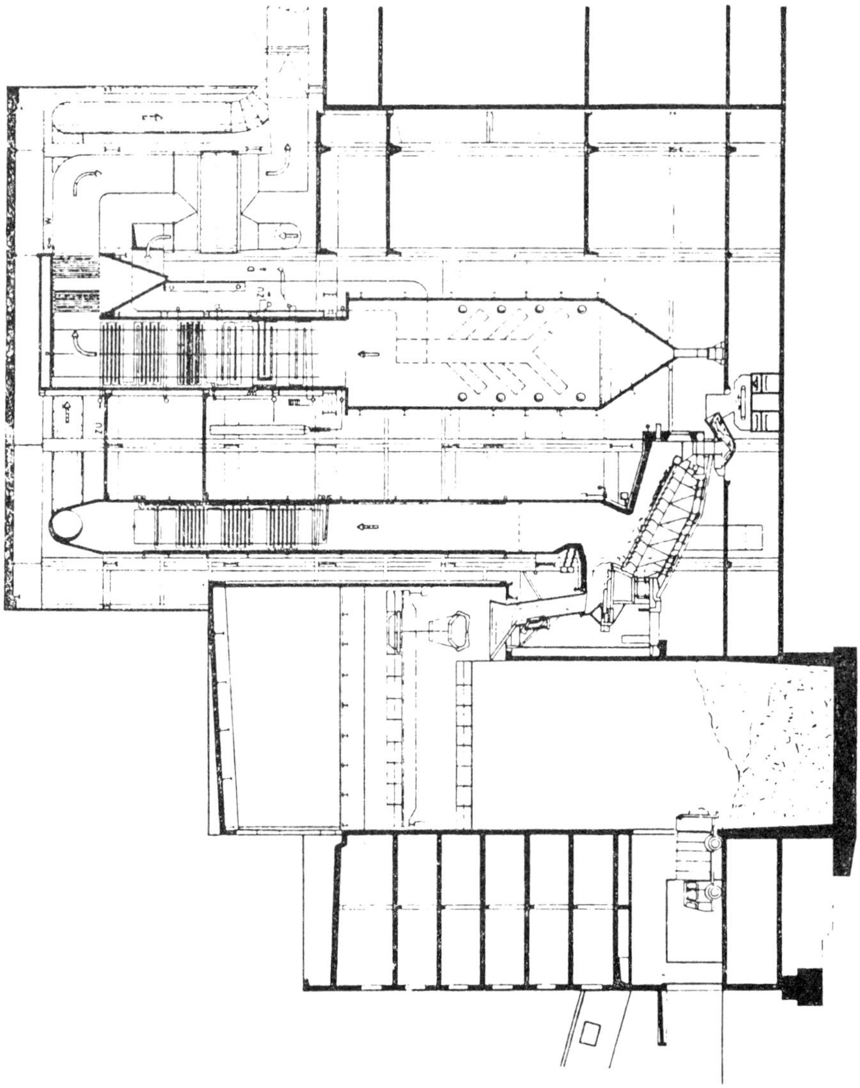

FIGURE 5.8: SECTION DRAWING OF MUNICH SOUTH IV AND V INCINERATORS

in the Munich North plant. The South units are again high pressure and temperature Benson boilers, but instead of having common evaporator, superheater, and reheater elements for both waste firing and fossil fuel firing, the South units completely separate the waste and fossil fuel furnaces. The boilers are coupled in such a way that the waste-fired unit serves as an economizer for the main boiler of the fossil fuel-fired furnace. If the waste-fired boiler is shut down, the fossil fuel-fired unit can operate at 100% output capacity on its own.

The waste bunker and the loading crane layout is similar to that at the North plant. However, because land was scarce, the bunker is rather small for the capacity of the two furnaces. The waste feeding mechanism is also different from the installations in the North. The South units have a double action horizontal piston pusher for coarse and fine control of waste volume being fed to the grate. The speed of the two pistons can be varied to regulate the input.

The grate is exactly the same size as in North II, and also consists of five parallel runs. The air inlet system also follows the same basic principle, having six air inlet regions down the length of the grate, each with a regulated amount of air against a 300 mm back pressure to insure uniform burning across the grate. The stoker movement is hydraulic rather than mechanical.

As in all of the other furnaces, the waterwalls do not extend all the way down, but start about 10 meters (33 ft) above the grate area. This causes intense heat in the lowest portion of the firebox and limits the amount of waste that can be burned at any one time to less than the designed capacity.

The waste-fired furnace consists of a single vertical unit with a gas exit at the top leading directly to the air filter and to the stack. The fossil fuel furnace also is vertical with a subsequent set of tubes in a horizontal passage through which the gases pass before continuing to the electrostatic precipitator and to the common exit up the stack. The South plant can burn either natural gas or oil.

The condenser cooling water comes from the Isar Canal which runs right next to the plant at the same elevation. Each of the South units (like North II) has three Martin ash quenchers. The ash is conveyed by belt to a processing plant where ferrous material is recovered. The remaining ash is then stored in another bunker prior to being hauled away by truck to the city dump. The turbine units are rated at 124 MW (155 MW at 0.8 cosine phi).

The South V unit entered operation in April 1969. The South IV unit followed in August 1971. The South IV is essentially the same as South V except that there is a bigger economizer in the waste-fired furnace and there are more economizer tubes in the fossil fuel-fired furnace.

The Basic Steam and Feedwater Flow System

Figure 5.9 shows the basic flow system for steam and boiler feedwater in the

FIGURE 5.9: STEAM AND FEEDWATER FLOW SYSTEM FOR MUNICH SOUTH IV AND V INCINERATORS

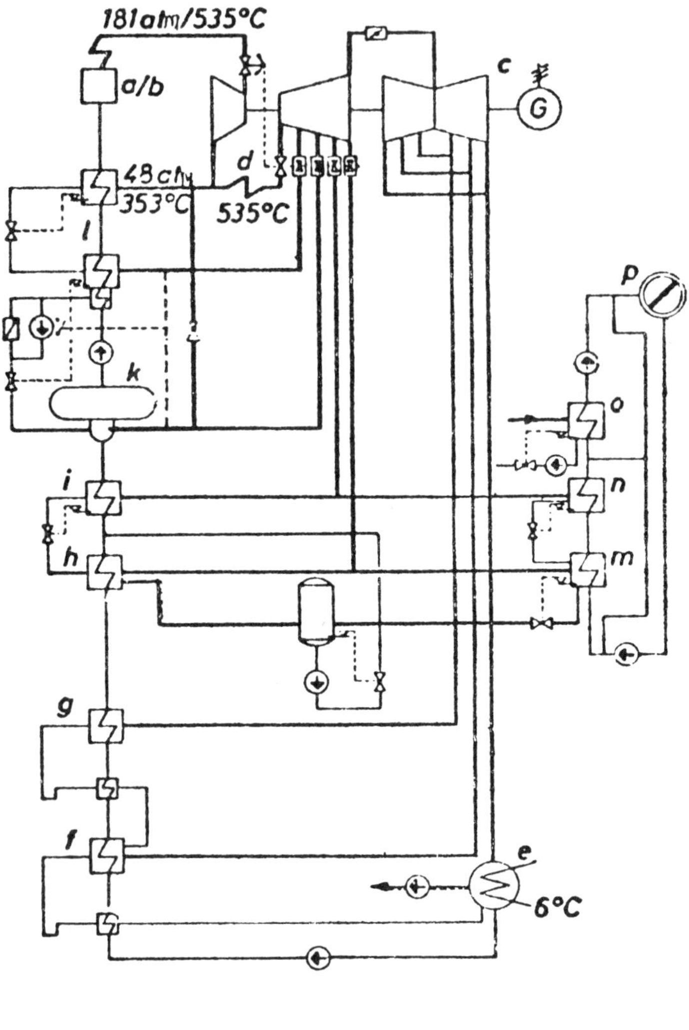

a. Boiler	i. Low pressure heater
b. Boiler	k. Feedwater tank and deaerator
c. Turbo-generator	l. High pressure heater
d. Reheating	m. District heating heat exchanger
e. Main condenser	n. District heating heat exchanger
f. Vacuum preheater	o. Nozzle preheater
g. Low pressure heater	p. District heating net (Perlach)
h. Low pressure heater	

Munich South IV and V combined waste and gas units. The Benson boiler is split into two parts: the waste portion fires an economizer and the feedwater is subsequently transformed into superheated steam in the gas fired section. This steam is delivered to a three part turbine. After the first pass through the initial turbine unit, the exhaust steam is reheated before passing through the two subsequent stages. Steam is tapped off a total of seven times for various heating functions, before the remaining steam is condensed in the main condenser.

The relationship between the two boilers (waste-fired and natural gas-fired) as well as the valve circuitry and the 150% redundancy of the feedwater pumps is shown in Figure 5.10. Also shown is the reheat phase with steam at 48.6 bars (48 atm) and 353°C (667°F) at reheater entry and at 46.6 bars (46 atm) and 540°C (1004°F) at the reheater outlet.

Level of Automation

The Munich South incinerator units are as highly automated as the others. The loading hoppers feed refuse down the chute to the furnace input feeding sytem. This system automatically regulates the desired flow of waste to the top of the waste burner. This system is more sophisticated than that on the North units and allows a more precise control of the waste inlet volume.

The air supply system under the grate is also similar to that of the North units. The hydraulically activated stoker is automatically operated from the control room to adjust for furnace gas temperature and economizer outlet temperature. The ash roller speed is also automatically adjusted from the control room.

Ash quenching and evacuation are fully automatic. Ash drops from the ash roller into a shallow trough of water which is constantly refilled. It is then shoved up an incline, and tumbled through a rough grate (to remove bulky items) onto a continuous moving belt. This belt brings the ash to a separate building where ferrous metal is automatically removed and loaded onto freight cars. The remaining ash is conveyed to a bunker to await final trucking to the dump.

All of the critical valves are remotely operable from the main control room with back-up manual capability. The German boiler code requirements for redundancy of valves and other critical elements such as feedwater pumps are the same as for the other units. Other redundant elements include a parallel conveyor belt for ash evacuation.

As in the case of the North units, however, there is no crossover capability for steam from one boiler to the other turbine. This is not possible due to the complexity of the reheating systems and the multiple stages of drawn off steam.

FIGURE 5.10: HEAT FLOW IN MUNICH SOUTH IV AND V INCINERATORS

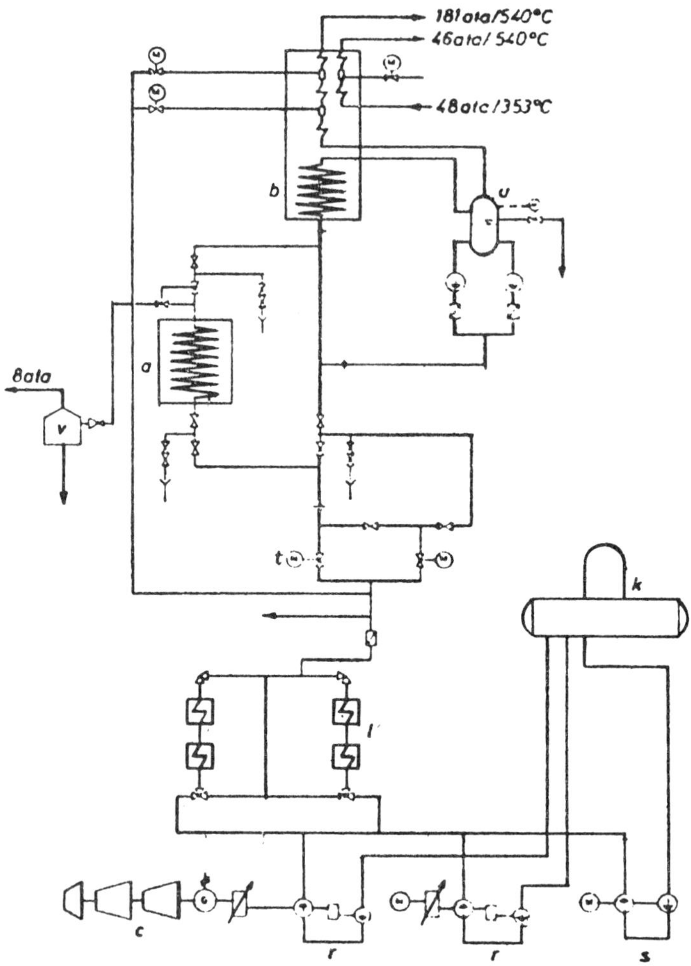

a. Refuse boiler
b. Main boiler
c. Turbo-generator
k. Main feedwater tank with deaerator
l. High pressure preheater
r. Full load feed pumps
s. Half load feed pump
t. Motor valve
u. Start-up drum
v. Relief unit

System Performance History

Table 5.15 shows the annual operating characteristics of the South IV and V units. From the beginning of operations through the end of 1975, a total of 718,269 metric tons (792,220 short tons) of waste was burned in South V and 467,351 metric tons (514,817 short tons) of waste were burned in South IV.

The average heat value of the waste incinerated in the two waste-fired furnaces has increased from 1,700 kcal/kg (3,060 Btu/lb) to over 1,850 kcal/kg (3,330 Btu/lb).

By the end of 1975, the South V waste-fired furnace had accumulated a total of 31,539 hours in service, while the newer South IV unit had amassed 22,038 hours. The maximum amount that either unit has ever operated in any one year was 66% of the total available time, in 1975 for South IV.

The rate of incineration has varied from year to year and from one furnace to the other. In 1975, the average burning rate was 21.3 metric tons (23.5 short tons) per hour of operation of South IV and 25.1 metric tons (27.6 short tons) per hour of operation of South V. The designed capacity was 40 metric tons (44 short tons) per hour per furnace (at 1,500 kcal/kg or 2,700 Btu/lb).

Materials and Energy Balance Calculations

The daily energy balance for South IV and V (see Table 5.19) is calculated in the same manner as the daily energy balances for the North plants.

Tables 5.16 and 5.17 show the most recent annual energy balance for South IV and V. The average heat input from waste was 15.1% for the IV unit in 1975, and 15.8% for the V unit. The designed maximum heat contribution for these units was 20%.

On the average, the units produced one kWh of electricity for a heat input (waste and gas) of 2,363 kcal or 9,377 Btu (unit IV) and 2,568 kcal or 10,191 Btu (unit V). If gas alone had been used, the kcal values required for each plant would have been 2,282 and 2,421 (9,050 and 9,607 Btu). Waste alone would have required 2,969 and 3,870 kcal (11,782 and 15,358 Btu).

Based on a conversion factor of 860 kcal/kWh, the overall efficiencies of each unit in 1975 in terms of heat transformation into electricity were:

	Unit IV, %	Unit V, %
Average (mixed fuels)	36.4	33.5
Fossil Fuel Only	37.7	35.5
Waste Only	29.0	22.2

TABLE 5.15: ANNUAL OPERATING CHARACTERISTICS OF MUNICH SOUTH INCINERATORS IV AND V

Year	Waste Burned (metric tons)*		Average Heat Value (kcal/kg)**		Hours in Service				Average Metric Tons Per Hour	
	IV	V	IV	V	IV	%	V	%	IV	V
1969		58,859		1,710			2,512	29		23.4
1970		121,770		1,682			4,630	53		26.4
1971	22,691	117,617	1,766	1,746	1,272	15	4,994	57	17.8	23.6
1972	106,705	76,170	1,773	1,765	5,404	62	4,919	56	19.8	15.5
1973	115,905	104,942	1,814	1,830	5,201	59	4,921	56	22.3	21.3
1974	99,402	118,009	1,797	1,869	4,407	50	4,739	54	22.6	24.9
1975	122,648	120,902	1,844	1,865	5,754	66	4,824	55	21.3	25.1
Total	**467,351**	**718,269**			**22,038**		**31,539**		**21.2**	**22.8**

* To convert metric tons to short tons, multiply by 1.1.
**To convert kcal/kg to Btu/lb multiply by 1.8.

TABLE 5.16: ENERGY BALANCE FOR THE MUNICH SOUTH V INCINERATOR IN 1975

Inputs	Metric System	English Equivalent
Coal	None	
Natural gas		
Amount (volume)	134,559.3 thousand m^3	4751 mmcf
Heat value	8920 kcal/m^3	1003 Btu/cu ft
Total heat	1,200,249 Gcal	4.76×10^{12} Btu
% total heat in	84.1	
Oil (heavy)		
Amount consumed	125 metric tons	137.7 short tons
Heat value	9904 kcal/kg	17,828 Btu/lb
Total heat	1238 Gcal	4.91×10^{9} Btu
% total heat in	0.1	
Waste		
Amount consumed	120,902 metric tons	133,181 short tons
Heat value	1865 kcal/kg	3357 Btu/lb
Total heat	225,504 Gcal	8.95×10^{11} Btu
% total heat in	15.1	
Total heat input		

Outputs	Metric System	English Equivalent
Electrical energy out		
Total produced	555,790 MWh	
Total delivered	530,507 MWh	
Self-consumed energy	25,283 MWh	
Self-consumed heat	4122 Gcal	1.64×10^{10} Btu
Heat in electricity and self usage	490,010 Gcal	1.94×10^{12} Btu
Total heat as % of total heat in	34.3	
District heat energy out (steam)	130,749 Gcal	5.19×10^{11} Btu
% total heat in	9.2	
Boiler losses (% heat in)	10.6	
Startup/shutdown losses (% heat in)	0.3	
Condensor and other losses (% heat in)	45.5	
Total losses (% heat in)	56.4	
Total losses	805,591 Gcal	3.19×10^{12} Btu
Specific heat required for electricity production		
Fossil fuel only	2421 kcal/kWh	9608 Btu/kWh
Waste only	2969 kcal/kWh	11,782 Btu/kWh

TABLE 5.17: ENERGY BALANCE FOR THE MUNICH SOUTH IV INCINERATOR IN 1975

Inputs	Metric System	English Equivalent
Coal	None	
Natural gas		
Amount (volume)	142,538.6 thousand m^3	5033 mmcf
Heat value	8914 kcal/m^3	1002 Btu/cu ft
Total heat	1,270,631 Gcal	5.04 x 10^{12} Btu
% total heat in	84.8	
Oil (heavy)		
Amount consumed	85.3 metric tons	93.9 short tons
Heat value	9894 kcal/kg	17,810 Btu/lb
Total heat	844 Gcal	3.35 x 10^9 Btu
% total heat in	0.1	
Waste		
Amount consumed	122,648 metric tons	135,105 short tons
Heat value	1844 kcal/kg	3319 Btu/lb
Total heat	266,138 Gcal	1.06 x 10^{12} Btu
% total heat in	15.1	
Total heat input	**1,497,613 Gcal**	**5.94 x 10^{12} Btu**

Outputs	Metric System	English Equivalent
Electrical energy out		
Total produced	633,780 MWh	
Total delivered	605,331 MWh	
Self-consumed energy	28,449 MWh	
Self-consumed heat	12,117 Gcal	4.81 x 10^{10} Btu
Heat in electricity and self usage	558,562 Gcal	2.22 x 10^{12} Btu
Total heat as % of total heat in	37.3	
District heat energy out (steam)	139,940 Gcal	5.55 x 10^{11} Btu
% total heat in	9.4	
Boiler losses (% heat in)	9.4	
Startup/shutdown losses (% heat in)	0.4	
Condensor and other losses (% heat in)	43.5	
Total losses (% heat in)	53.3	
Total losses	798,202 Gcal	3.17 x 10^{12} Btu
Specific heat required for electricity production		
Fossil fuel only	2282 kcal/kWh	9056 Btu/kWh
Waste only	2969 kcal/kWh	11,782 Btu/kWh

Maintenance History, Repairs and Modifications

Unlike the North I furnaces, the waste-fired furnaces in the two South incinerators have not shown major corrosion problems as had been originally feared. The major maintenance of these units is performed during an annual plant stoppage of a month to five weeks. During this period the grate, the waste-fired boiler, and the other elements are checked, repaired or replaced.

More frequent maintenance is possible because of the lack of waste to burn on weekends. The waste bunker barely holds a two-day supply, whereas the collection service weekend effectively runs from Friday afternoon to late Monday morning. Thus each waste incinerator can be shutdown every other weekend. During these periods they are inspected and minor repairs can be made, thereby preventing more serious problems.

This design, where the waste-fired boiler serves as an economizer for the main boiler of the fossil fuel-fired furnace, has caused some operating problems, however. Sudden variations in the heat intensity of the waste-fired furnace can lead to premature evaporation of the economizer water. This in turn causes steam and water to spurt into the economizer of the main boiler disrupting the evaporation phase of this boiler thereby temporarily lowering the steam temperature. The operating rate of the turbine must then be adjusted.

In the meantime heat of the waste-fired furnace has to be lowered, either by reducing feed or by reducing underfire air. This procedure refills the tubes of the economizer but starves the main boiler of feedwater temporarily, and again the main boiler operating conditions must be adjusted to prevent overheating of the steam.

Innovations have been introduced into the automatic compensation of heat fluctuations in the waste-fired boiler. Temperature gauges in the boiler provide a direct input to the system controlling rate of feed of waste to the grate. Higher temperatures slow the rate of the double piston feeding mechanism, and also the rate of action of the grate elements. This change has been credited with reducing possible temperature fluctuations from 10% to 5%.

No major modifications have been made to either of the South units, but South IV (which was built two years after South V) is slightly different. More economizer tubes were added to the waste-fired boiler to decrease the flue gas exit temperature which was found to be too high. The economizer section of the main boiler was also redesigned.

The lowest portions of these incinerators are not cooled by waterwalls, but have a refractory lining (as do all the Munich units) up to a height of about 10 meters (33 ft) above the waste grate. In addition, the furnace narrows at the point where the waterwalls begin, and this shape, combined with the refractory lining, leads to much more intense heat in the grate area than seems necessary, given the higher heat value of the waste now being burned. The general absence of corrosion in these boilers speaks well of their design and demonstrates the wisdom of having the lowest tubes protected with a ceramic coating from the beginning.

Table 5.18 breaks the available hours for each waste-fired boiler in 1975 into operating time, scheduled maintenance, unscheduled maintenance and reserve time. Planned maintenance accounted for 8% of the total annual hours for each unit in 1975. Unplanned stoppages accounted for 2 and 3%, respectively. The units operated for 66% and 55% of the time, leaving 25% and 33% as reserve time. Combined operating and reserve hours show an availability of 90 and 89%.

The South IV main boiler was available 92% of the time (or 2% more than the waste boiler) whereas the South V main boiler was available only 83.5% of the time vs 89% for the waste boiler. The large amount of reserve time is due to two facts, the lack of waste to burn and the demand characteristics for electricity in the Munich area.

The waste boilers were designed to produce 167 GJ (1.58 × 10^5 Btu) per hour. In 1975, the two South waste boilers provided heat at an average rate of only 121.2 and 119.7 GJ, an average of 72.6% and 71.7% of their load capacities. Multiplying those figures by the percentage of time these plants were actually in operation gives the effective output of 47.7% of total capacity for South IV and 39.5% of total capacity for South V.

The output from the main boilers of these same plants in 1975 was 64.1% and 57.1%. Not only were the main boilers in use more often, but they also ran at a higher percentage of their designed capacity when in use.

Planned maintenance is, as usual, scheduled around institutional and equipment requirements. An annual period for grate and other component repair and replacement also necessitates a shutdown of these furnaces for about 4-5 weeks. As mentioned earlier, frequent routine maintenance is possible in each furnace every two weeks due to shutdown for lack of waste to burn on weekends.

Table 5.19 shows the average values of the South IV and V waste and main boilers and electricity generation plants on a typical day (Sept. 28th) in 1976. On that day, both waste and main boilers were in service for the entire 24 hour period.

The average steam rates were 306 metric tons (337 short tons) per hour on South IV and 281 metric tons (310 short tons) per hour on South V. This was at 84% capacity on South IV and 77% on South V. Feedwater temperature was 3°C cooler on South IV than on South V, and the waste flue gas temperature was only 202°C (396°F) on South IV whereas it was 307°C (585°F) on South V.

Waste was burned at a rate of 24 metric tons (26.4 short tons) per hour on South IV and 28 metric tons (30.8 short tons) per hour on South V. This was at

TABLE 5.18: AVAILABILITY OF MUNICH SOUTH INCINERATORS IV AND V IN 1975 BY MONTH (hours)

Month	Hours in Service		Planned Stoppages: Repair and Maintenance		Unplanned Stoppages		Reserve Hours		Total
	IV	V	IV	V	IV	V	IV	V	
January	662.0	73.5					82.0	670.5	**744**
February	455.5	250.0					216.5	422.0	**672**
March	491.0	508.5					253.0	235.5	**744**
April	542.0	462.0				1	178.0	257.0	**720**
May	410.5	548.5			43	2	290.5	193.5	**744**
June	478.0	522.0			7	15	235.5	183.5	**720**
July	650.5	418.0			10	97	83.0	179.0	**744**
August	28.5	544.4	690	48		66	25.5	85.5	**744**
September	692.0	3.0		696	9		19.0	21.0	**720**
October	418.0	637.0				1	326.0	106.0	**744**
November	426.0	442.5			76	38	218.0	239.0	**720**
December	500.0	364.0				63	244.0	317.0	**744**
Total	**5,754.0**	**4,823.5**	**690**	**744**	**145**	**283**	**2,171.0**	**2,909.5**	**8,760**
Percent	**66**	**55**	**8**	**8**	**2**	**3**	**25**	**33**	**100**

TABLE 5.19: DAILY OPERATING REPORT FOR MUNICH SOUTH INCINERATORS IV AND V (Sept. 28, 1976)

Boiler/Incinerator	Unit	MK-IV	K-IV	Total	MK-V	K-V	Total
Feedwater volume	metric tons	–	7,338.0	–	–	6,739.0	–
Feedwater temperature	°C	242.0	242.0	–	245.0	245.0	–
Raw steam volume	metric tons	–	7,338.0	–	–	6,739.0	–
Raw steam pressure	bar	206.0	177.0	–	220.0	177.0	–
Raw steam temperature	°C	335.0	530.0	–	340.0	530.0	–
Recovered heat/kg	kJ/kg	–	1,899.5	–	–	1,856.5	–
Total heat in steam	GJ	–	13,938.5	–	–	12,511.0	–
Reheating							
Volume	metric tons	–	7,026.6	–	–	6,598.2	–
Pressure-in	bar	–	37.0	–	–	36.0	–
Temperature-in	°C	–	330.0	–	–	330.0	–
Cooling spray	metric tons	–	314.0	–	–	485.0	–
Soot blowing	metric tons	–	20.0	–	–	20.0	–
Pressure-out	bar	–	34.0	–	–	33.0	–
Temperature-out	°C	–	530.0	–	–	530.0	–
Recovered heat (reheat) per kg	kJ/kg	–	568.2	–	–	637.0	–
Heat in received steam	GJ	–	3,992.2	–	–	3,992.2	–
Total heat in steam	GJ	3,060.0	17,930.8	20,990.8	3,007.0	16,714.3	19,721.3
Flue gas temperature after air preheater	°C	202.0	132.0	–	307.0	120.0	–
O_2 content after aph	%	12.0	4.0	–	11.0	4.0	–
Air temperature before aph	°C	45.0	45.0	–	40.0	40.0	–
Boiler losses in–							
Flue gas	%	21.6	4.4	–	33.1	4.0	–
Ash heat	%	1.4	–	–	1.4	–	–
Unburned ash	%	–	–	–	–	–	–

(continued)

TABLE 5.19: (continued)

Boiler/Incinerator	Unit	MK-IV	K-IV	Total	MK-V	K-V	Total
Unburned flue ash	%	–	–	–	–	–	–
Radiation losses	%	2.3	1.2	–	2.3	1.3	–
Boiler efficiency	%	74.6	94.3	90.8	63.1	94.6	87.9
Natural gas usage	KCBM I.N	–	509.6	–	–	473.6	–
Heat	GJ	–	19,006.5	19,006.5	–	17,666.9	17,666.9
Waste tonnage	metric tons	577.4	–	–	670.9	–	–
Heat (7,100 kJ/kg)	GJ	4,099.3	–	4,099.3	4,763.7	–	4,763.7
Total fuel heat	GJ	4,099.3	19,006.5	23,105.8	4,763.7	17,666.9	22,430.6
Hours in service	hours	24.0	24.0	48.0	24.0	24.0	48.0

Turbine	Unit	T-IV	T-V
Steam received	metric tons	7,338.0	6,739.0
Live steam pressure	bar	177.0	177.0
Live steam temperature	°C	530.0	530.0
Specific steam consumed	kg/kWh	3.1	3.0

Generator	Unit	G-IV	G-V
Electricity produced	MWh	2,240.0	2,260.0
Electricity consumed	MWh	102.0	71.0
Electricity delivered	MWh	2,138.0	2,189.0
Maximum power output	MW	116.0	120.0
Average power output	MW	97.6	94.2
Specific heat consumed	kJ/kWh	10,479.8	10,247.0

Source: Munich Electricity Works.

60% of capacity for IV and 68% for V (at 1,500 kcal or 5,952 Btu) based on a capacity of 40 metric tons or 44 short tons per hour.

The electrical outputs from each plant during the 24 hour period were 2,240 MW and 2,260 MW, respectively. Maximum power output was 116 MW from South IV and 120 MW from South V while average was 97.6 MW for South IV and 94.2 MW for South V. Estimated specific heat consumption was 10,479 kJ/kWh for South IV and 10,247 kJ/kWh for South V, showing marginally better performance from V than from IV.

OTHER COMMON SYSTEM FEATURES

Characteristics of Recovered Energy and Materials

The main form of energy recovered at the Munich North and South waste and fossil fuel power plants is electricity which is generated at 50 cycles, three phase, 10,500 volts. This comes from the turbo-generator units of 68 MW, 120 MW and 2 × 125 MW.

At the North plants, energy is also recovered as steam at 120°C (248°F) and used for district heating. At the South plant, energy is recovered both as hot water and as steam for district heating. In the South plant the hot water is at 125°C (257°F) returning to 50°C (122°F). A maximum of 80 metric tons (88 short tons) of steam per hour can be furnished to the city network.

Ferrous materials are separated electromagnetically from all furnace ash. These are shipped to a scrap metal dealer who has a concession from the Electricity Works. This scrap is very dirty and of low quality. It amounts to 1.6% by volume of total waste burned in the North plants in 1975 and 2% in the South plants. Sometimes the scrap is baled prior to shipment and sometimes it is loaded in bulk. The remaining ash is hauled to the dump and serves merely to cover other unburned waste.

The Market for Recovered Energy and Scrap Iron

The electrical energy produced in part by the burning of waste is sold by the city Electricity Works to private, public, and commercial and industrial users. The present tariffs charged for electricity in the city of Munich are shown in Table 5.20.

The charge for district heating is based on three distinct elements:

A basic charge of 21 DM for each 1,000 kcal (3,968,500 Btu) per hour of the user's basic capacity.

A price of 35.50 DM for each Gcal (3,968,500,000 Btu) of heat supplied.

Rent for the meter at the following rates:

Meter Size	Yearly Fee
10m /hr	240 DM
10-15m /hr	324 DM
over 25m /hr	420 DM

TABLE 5.20: ELECTRICITY RATES

	Tariff I	Tariff II
Regular price* (DM/kWh)	**0.135**	**0.104**
Additional charges added to regular price (DM)		
Households		
For first 2 rooms	5.00	9.65
For each extra room	1.25	2.15
For heavy appliance usage: each 0.5 kW up to 2kW	5.50	11.10
each 0.5 kW over 2 kW	5.50	11.10
Farms		
For first 5 hectares	4.50	8.50
For each additional hectare	0.95	1.35
Advertising, professional, and other needs		
Lighting (each 100 watts)	2.75	4.20
Other (each 500 watts)	5.50	11.10

NOTE: In 1976, 2.52 DM = $1.00.

*Special off-peak rates (for Tariff I or II users only): 0.078 DM/kWh.
Small users price (cheapest up to 174 kWh/year for 2-room unit): 0.48 DM/kWh.

The city of Munich has a contract with a private scrap metal dealer in Munich who buys the scrap by the train car load. The price received by the Electricity Works is reviewed about every 3 months, with pressure for revision by the dealer if the price for scrap is falling and by the Electricity Works if it is rising.

The 1976 price was about 20 DM per metric ton (18 DM per short ton) FOB incinerators. The Electricity Works pays demeurrage charges to the railroad at a rate of:

1st day	11.50 DM
2nd day	30.00 DM
3rd day	64.00 DM
4th day	114.00 DM

There is thus a premium on filling the railroad cars as efficiently as possible.

Total revenues from scrap metal amounted for each plant to the following sums in 1975:

North IA	11,548 DM
North IB	9,719 DM
North II	44,308 DM
South IV	49,059 DM
South V	48,360 DM

Environmental Control Systems

West German laws set environmental standards for both the emission of pollution into the air, and the quality of the air breathed by living beings.

The law covering emissions in effect when the Munich North and South plants were built was the TA Luft 1964 (Technischen Anleitung zur Reinhaltung der Luft) issued by the Bundesminister des Innern (Federal Minister of the Interior).The standard which then applied to the plants set a limit of 150 mg of particulates per cubic meter of air, for volumes of 100,000 m^3 per hour. For solid fuels, the limit was set at a certain percentage of the total amount of ash generated at full operation. The law also limited fine particulate matter (less than 10 microns) to 150 mg/m^3. There were no other limits on emissions from plants.

A new law was passed in 1974 which lowers the particulate limit for such plants to 100 mg/m^3 for fine particles and total ash. It also sets limits on CO, NO_2 and SO_2. These standards are not applicable to the present units, however, and no retrofitting is required.

The law concerning air quality specified limits to the amount of particulates deposited per m^2, both for short and long term periods. Limits were also set for the quantities of various gases in the average m^3 of ambient air for short and longer term periods.

In 1964 the short term particulate limit for nontoxic residues was 1.3 g/m^2. The longer term average limit was 0.85 g/m^2 per day. These limits were reduced in 1974 to a short term limit of 0.65 g/m^2 and a long term average limit of 0.35 g/m^2. Standards have been set for fine particulate matter (less than 10 microns) of 0.2-0.4 for short term and 0.1-0.2 on a longer term average.

For gases, the 1964 short-term limits were 2 mg/m^3 for NO_2, 0.6 mg/m^3 for Cl_2, 0.75 mg/m^3 for SO_2, and 0.3 mg/m^3 for H_2S. The longer term average limits were 1 mg/m^3 for NO_2, 0.3 mg/m^3 for Cl_2, 0.4 mg/m^3 for SO_2, and 0.15 mg/m^3 for H_2S.

The new short-term limits are 0.3 mg/m^3 NO_2, 0.6 mg/m^3 NO, 0.3 mg/m^3 Cl_2, 0.2 mg/m^3 inorganic gaseous form of Cl ions, 0.004 mg/m^3 inorganic gaseous form of F ions, 30 mg/m^3 CO, 0.4 mg/m^3 SO_2 and 0.01 mg/m^3 H_2S. The new limits for longer term averages are 0.1 mg/m^3 NO_2, 0.2 mg/m^3 NO, 0.1 mg/m^3 Cl_2, 0.1 mg/m^3 inorganic gaseous Cl ions, 0.002 mg/m^3 inorganic gaseous F ions, 10 mg/m^3 CO, 0.14 mg/m^3 SO_2, and 0.005 mg/m^3 H_2S.

There are no water pollution problems from any of the Munich incinerators because the Martin ash quenchers do not recirculate water. They provide a continuously refilling bath into which the ash drops. After being quenched, the ash is forced out to the discharge conveyors. The water is consumed either as steam which rises into the furnace or as the moisture content of the quenched ash (an average of 15%).

The only air pollution controls on these furnaces are Lurgi electrostatic precipitators designed to reduce particulate matter by 99.5-99.7%. The North plants originally had 80 meter (262 ft) stacks, but these have been raised to 130 meters (427 ft). The South plants share a common 130 meter (427 ft) stack.

Aesthetics and Layout

The general layout of the Munich North Plants I and II is shown in Figure 5.11. Collection trucks have easy access to the plants.

The general layout of the Munich South incinerators IV and V is shown in Figure 5.12. This installation was built on a narrow strip of land bounded on one side by the Isar Canal, on another by a series of railroad tracks, and on the third and fourth sides by an older heating plant and a highway. This restricted space accounts for the limited capacity of the waste bunker which can store only 4,000 m^3 of waste, barely enough to fire the plants for 2 days. When this plant was built, it was necessary to build a complicated underground passage under the highway to ease the traffic of collector vehicles.

ECONOMICS AND MANAGEMENT

Original Cost and Amortization

The original capital cost of the North and South plants and of the waste incineration portions is shown below. Information is only available on the total cost of the North and South plants in aggregate, hence for the North plant units IA, IB and II are combined, and for the South plant units IV and V are combined. In both cases, some of the costs have been allocated by an arbitrary distribution method. (See Tables 5.21 and 5.22.)

The entire North plant (I and II) cost 151.1 million DM ($37.8 million, 1960-1966 or $88.3 million, 1976). Of this the waste section is estimated to have cost 25.8 million DM ($6.45 million, 1960-1966 or $10.2 million, 1976). The entire South plant (IV and V) cost 140.7 million DM ($39.2 million, 1969-1971 or $67.9 million, 1976), of which the waste portion is estimated to have cost 35 million

FIGURE 5.11: PLAN OF MUNICH NORTH I AND II INCINERATORS

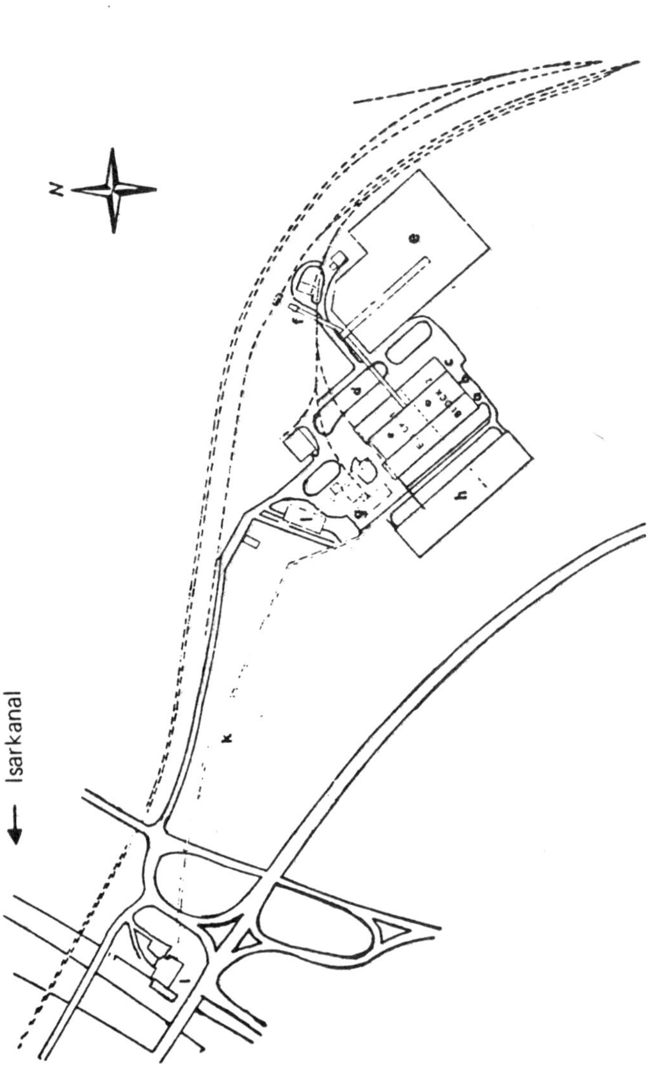

a. Machinery building
b. Boiler building
c. Waste bunker
d. Coal conveyor
e. Coal storage
f. Coal transfer system
g. Operations building
h. Open air transformers
i. Pumping station
k. Cooling water conduits
l. Ash loading building

FIGURE 5.12: PLAN OF MUNICH SOUTH IV AND V INCINERATORS

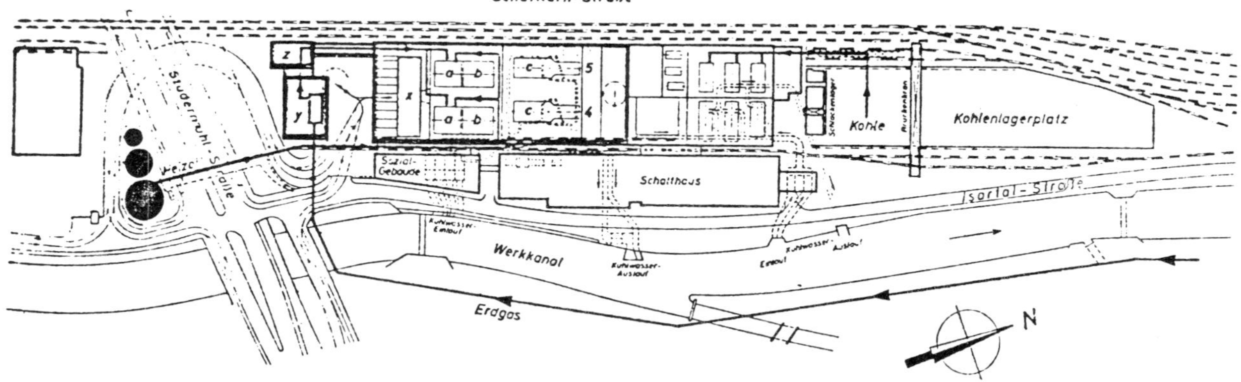

a. Waste boiler
b. Main boiler
c. Turbo-generators
x. Waste bunker (8m deep)
y. Ash building
z. Breaking tower

TABLE 5.21: MUNICH SOUTH INCINERATOR INSTALLATION
(Costs from Electricity Works Records of May 14, 1969, in million DM)

Total Original Capital Costs		**140.70**
Costs for the Waste Portion		**35.00**
Details of Waste Capital Costs		
Building and construction		
Boiler building (part cost)		2.00
Waste building and discharge stalls		4.10
Waste delivery box doors		0.31
Ash treatment buildings, etc.		1.78
Room for ash bunker		0.33
Partial cost of streets, etc.		0.10
Water drainage and related		0.28
Total construction		**8.90**
Machinery and equipment		
Total boiler cost	35.4	
Conventional boiler 365 tph (365 tph x 30,000 DM/t/hr) =	–22.0	
Additional cost for waste boiler	**13.4**	
Waste boiler		13.40
Electrostatic precipitator		2.40
Ash treatment		1.16
Tubing (partial cost)		0.80
Measurement and control instrumentation		0.60
Scrap press, railway capstan		0.43
Filter for waste building air		0.34
Air entry and exit from waste building		0.36
Painting and miscellaneous		0.67
Total machinery costs		**19.80**
Electrical installation		
Total electrical plant	17.5	
Less transformers, generator protection, etc.	–4.0	
Remainder representing self consumption of electricity	**13.5**	
Total plant consumption	24.0 MW	
Consumed by waste portion	4.4 MW	
Proportion of total cost for waste		**2.60**
General additional costs (partial cost)		**0.60**
Special street underpass	3.5	
Less 10% for other use	–0.4	
	3.1	**3.10**
Final total		**35.00**

NOTE: From 1969-1971, 3.59 DM = $1.00.

TABLE 5.22: MUNICH NORTH INCINERATOR INSTALLATION
(Costs from Electricity Works Records of April 7, 1967
in million DM)

Total Original Capital Cost	**151.100**
Costs for Waste Portion	**25.800**
Details of Waste Capital Costs	
Buildings and construction	
Waste building	4.600
Ash system (breaking tower, ash bunker, conveyor bridge, ash building)	0.610
Water evacuation (partial cost)	0.240
Main boiler building space	1.380
Total	**6.830**
Access roads (partial cost)	**1.000**
Machinery and equipment	
Cranes	1.150
Air inlet, conditioning	0.710
Truck weighbridge	0.028
Water evacuation pump (waste building)	0.045
Fire protection system (partial cost)	0.022
Extra cost for boiler system	
–Grates	3.007
–Grate accessories	1.970
–Tubing	1.173
–Larger electrostatic precipitator	0.350
–Measuring and control instruments	0.550
–Ash treatment system	1.155
–Ash extraction equipment (from bunker)	0.190
–Scrap metal recovery equipment	0.299
–Railway car movement capstan	0.009
–Painting and miscellaneous	0.195
Original total	**10.853**
Plant modification to steam/water separation system as decided on 2/29/72	0.850
Revised total	**11.703**
Electrical installation (partial cost)	**1.441**
General overall costs (groundwork, canalization, workshops, social buildings, lighting, etc.) (partial cost)	**4.876**
Final total	**25.850**

NOTE: From 1960-66, 4.00 DM = $1.00.

DM ($9.8 million, 1969-1971 or $13.5 million, 1976). For the North plant, average waste burning capacity costs amounted to 286,700 DM per metric ton ($65,000 per short ton, 1966) of waste per hour. For the South installation, a capacity of one metric ton per hour cost 437,500 DM ($110,300 for one short ton per hour, 1969-1971).

For the city's accounting purposes, an annual interest charge of 6.5% is assumed for unrepaid capital. The costs are amortized on three different schedules. The Munich South plant underpass is given 40 years. The mechanical and electrical installations are given 12 years and the remaining building and general expenses are given 30 years.

These three periods with the interest charge establish a constant annual annuity value of 7.07%, 7.66% and 12.26%. This creates an annual amortization charge buildup as follows:

	North Plant (1960-66)		South Plant (1969-71)		Total
	Capital 10^6 DM	Amort. DM	Capital 10^6 DM	Amort. DM	Amort. DM
7.07%	-		3.1	291,170	291,170
7.66%	12.71	973,586	9.5	727,700	1,701,286
12.26%	13.14	1,610,964	22.4	2,746,240	4,357,204
Total	25.85	2,584,550	35.0	3,693,110	6,277,660

Conversion:	DM/$
1960-66	4.00
1969-71	3.59

Operating Costs and Revenues

Every three years, the city of Munich's Electricity Works puts together a report of the anticipated costs and revenues for the treatment of waste and recovery of energy and materials from waste for the following three years. This report is the basis for attribution of financial credit to the Electricity Works on the city books.

The cost components for the projected period 1973-1975 were computed in January 1973 and served as the justification of the city's giving the Electricity Works a credit of 18.80 DM per metric ton ($6.39; 1973 per short ton) of waste to be treated during this subsequent period.

The cost information shown below is taken partly from the 1973 report and partly from more recent information obtained during interviews in Munich in October 1976. From these various bits of information it has been estimated approximate cost and income statements for the North and South waste treatment plants.

The 1973 Munich report was divided into seven major cost categories: a capital cost (amortization and interest), a constructed (fictitious) capital tax charge, a constructed (fictitious) industrial revenue tax, personnel costs, other operating costs, repair and maintenance costs, and general overhead costs. There were also two revenue categories, a payment by the Electricity Works for the value of heat recovered from the waste and payment received for scrap metal. Subtracting the revenues from the costs produced the figure 18.80 DM per metric ton ($6.39 per short ton) as the cost of treatment.

The capital annuity costs are shown in the above section. The constructed capital tax was assessed at 0.84% of the industrial capital invested. On an annuity basis, this comes to an annual charge of about 0.45% of the original capital costs or a total of 117,639 DM ($44,060, 1973) for the North plant, and 159,740 DM ($59,825, 1973) for the South plant.

The industrial revenue tax was calculated as 21% of the annuity interest charge. This varied in application to the three annuity periods from an effective rate of 0.82 to 0.96%. The total charge for the North plant was 223,409 DM ($83,670, 1973) and for the South plant 299,890 DM ($112,320, 1973). Personnel costs were estimated on the basis of the four categories of workers directly attached to the waste portions of each of the incinerator and power plants. The 1973 report showed worker groups and their annual salary levels as follows:

Employee Group	North		South		Total
	No.	Cost	No.	Cost	No.
V (23,534 DM)	18	423,612	19	447,146	37
IV (22,040 DM)	15	330,600	10	220,400	25
III (21,088 DM)	15	316,320	15	316,320	30
II (20,252 DM)	5	101,260	5	101,260	10
Total	53	1,171,792 ($438,870, 1973)	49	1,085,126 ($406,410, 1973)	102

These salaries were expected to increase at an annual rate of 8.5% during the three year period giving an average of 18% higher value for the median year. Thus the actual cost would be 1,382,715 DM ($533,870, 1974) at the North plant and 1,280,449 DM ($494,380, 1974) at the South plant.

It is interesting to note that not only at the individual plants, but also at the general operating staff level there is a distinct separation made between the technical operation and delivery of services on the one hand, and the financial and accounting activities on the other. The plants are run by the technical operators. Their costs are noted according to an elaborate accounting system which includes an attribution of worker hours and other expenses to various subunits of the plants.

The overall economic performance of these plants, however, is the concern of the top political and planning administrators rather than of the local technical or supervisory staff. For a variety of political reasons, costs are paid from and revenues are received into a number of different accounts. Attributed real costs and charged costs differ. Neither the electricity generation nor the waste burning functions are run exactly as individual cost and revenue centers.

As mentioned earlier, it is difficult to say exactly and unambiguously which operating personnel are used for waste burning and which are used for other firing and electricity generation. There are now about 233 employees in the North plant (including North IA and IB and North II units). Of this total, 160 are grouped in six shifts of about 26 men apiece. Three shifts work around the clock, a fourth serves as a rest period, a fifth is for reserve, and a sixth is for the accumulated extras of sickness, leave, etc.

The director of the North plant estimated the work force dealing specifically with waste as follows:

11 crane operators
11 skilled grate men
11 ash system tenders
17 general workers for cleaning, etc.

It is difficult to ascertain the salary levels of these workers in terms of total charges paid. However, considering take-home pay, social security charges, 13th month, etc., this could average nearly 30,000 DM ($11,900, 1976) each. This is in excess of the figure shown on the cost statement of the Electricity Works which determines the charges paid by the waste generators.

There are about 170 employees at the South plant, with shifts of 20 men apiece, of whom seven work directly with the waste. The total work force for waste at the South plant is estimated to be:

12 crane operators
6 men on the grates
12 men on the ash quenchers
12 men in the ash building
5 additional men for weighing of waste, etc.

A slightly different estimate gave the North plant 54 men on the waste side, eight per shift, distributed as follows:

12 crane operators
12 men on the grates
12 men on the ash quenchers
12 men in the ash building
6 additional men for weighing of waste, etc.

Thus, even among the technical operating personnel in the two plants, there are differences of opinion as to which workers and which jobs are actually attributable to the waste treatment function.

These personnel figures include no men in the central control room, although there are functions in the control room that deal specifically with incineration of waste. Also, no day personnel are attributed to general repair and other maintenance services.

It is assumed that most of the remaining personnel at each plant are repair and general maintenance workers. No information is available on the number of these workers specifically responsible for work on the waste portions at any of the plants.

One crosscheck of these estimations can be made by consulting the annual financial accounts of the individual plants where hours of work are attributed for each worker to specific parts of the plant when and as this work is performed. In 1975, such information was collected for each plant and that of the South Plant was made available.

These data showed that for four sections of the plant which clearly dealt with waste (the IV and V waste furnaces, the ash treatment section, and the refuse bunker/crane operation), the total operating costs including personnel and supplies were 6,480,492 DM ($2,634,350, 1975). This should be compared to a personnel cost estimation of 1,085,126 DM ($441,100, 1975). There is obviously a major discrepancy here. Cost increases for the North plant are assumed to have been proportional and to total 6,996,000 DM ($2,843,900, 1975).

Other operating costs were estimated to be 50,000 DM ($20,325, 1975) each for the North and South plants. These costs (operating supplies, grease, etc.) are included in the total operating costs counted in the paragraph above.

Repair and maintenance costs for the two plants (including labor and materials) were reported to have increased as follows:

Evolution of Repair and Maintenance Costs (DM)

Year	North	South	Total	DM/$
1964	91,825		91,825	3.98
1965	264,290		264,290	4.00
1966	338,415		338,415	3.98

(continued)

Year	North	South	Total	DM/$
1967	513,621		513,621	4.00
1968	1,186,293		1,186,293	4.00
1969	1,044,075	165,617	1,209,629	3.69
1970	2,467,776	462,346	2,930,125	3.64
1971	3,296,949	782,848	4,079,798	3.48
....	....	....	....	....
1975			(estimated)$6,000,000	2.46

The combined costs were anticipated to rise from 4 to 6 million DM from 1972 to 1975, and an average of 5 million per year was estimated for this category.

A check on the accuracy of the repair estimates from the actual accounting report for 1975 shows that repair costs in the North plant were 2,443,700 DM, and in the South plant, 1,578,438 DM, for a total of 4,022,138 DM ($1,635,000, 1975). This is less than the forecast. (It is possible that the figures for the North plant were understated.)

The final cost category is the attribution to waste incineration of a part of the total general overhead costs up to and including local and general management of the entire Electricity Works administration as well as the general services, service vehicle operation, chemical laboratory analysis, repair workshops and all of the other nondirectly identifiable costs. Total general overhead costs for the North and South plants grew as follows from 1969-1971:

Year	North	South	Total	DM/$
1969	5,238,010	2,862,655	8,100,666	3.69
1970	5,544,959	3,826,177	9,371,137	3.64
1971	6,126,492	4,639,295	10,765,787	3.48

The 1973 report showed that according to an internal Electricity Works accounting formula, the overhead burden for the waste portion of the two plants combined should have been:

1969	1,388,700 DM
1970	1,968,300 DM
1971	2,653,840 DM

However, after discussions between the revisions bureau and the Electricity Works, the figure was downgraded to 2,358,000 for 1971 and, on the basis of conservative growth estimates, was fixed at an annual level of 2.5 million DM for the succeeding three years. A crosscheck for this figure is given by the amount shown in the 1975 accounts for the South plant alone where the total came to 2,672,920 DM ($1,086,550, 1975). This is greater than was forecast for both plants during that year. The general overhead at the North plant was 32% higher than at the South plant in 1971. Using this same ratio for 1975 would give a North overhead figure of 3,528,254 DM ($1,434,250, 1975) for 1975. Therefore, revised estimates for three cost categories, Personnel and Operating Costs, Repair Costs, and General Overhead, were made.

The original and revised cost accounts for 1975 for the North and South plants are therefore as follows:

Estimated 1975 Costs for Munich North and South Refuse Plants

(In thousand DM)

CATEGORY	NORTH Original	NORTH Revised	SOUTH Original	SOUTH Revised
Capital Annuity	2,585	2,585	3,693	3,693
Capital Tax	118	118	160	160
Revenue Tax	223	223	300	300
Personnel Costs	1,382	6,996	1,280	6,480
Operating Costs	50		50	
Repair Costs	3,500	2,443	1,500	1,578
Overhead	1,423	3,528	1,077	2,672
Total Costs	9,281	15,893	8,060	14,883
Cost per metric ton of waste burned	45.29 DM	77.55 DM	33.09 DM	61.11 DM

Revenue for the two plants in 1975 came from the sale of heat recovered from the waste and from the sale of scrap ferrous metal recovered from the ash. The 1973 report estimated that at the North plant 1.3 Gcal of heat was delivered for each metric ton of waste burned (4,690 Btu per short ton), while at the South plant 1.13 Gcal was delivered per metric ton (4,077 Btu per short ton). Estimates of likely tonnage were used to calculate the total heat likely to be delivered in the following years.

The recent average cost price per Gcal from coal and natural gas was used to set a value on the heat produced from waste. For coal, the cost was estimated at 15.50 DM per Gcal ($1.46, 1973, per 10^6 Btu) and for natural gas at 13.50 DM per Gcal ($1.27, 1973, per 10^6 Btu). The amount of waste to be burned each year at each plant was forecast at 237,500 metric tons (261,621 short tons). Thus the North plant would earn 4,785,625 DM ($1,792,370, 1973) per year, while the South plant, recovering less heat and having a lower fuel competition price, would only earn 3,623,062 DM ($1,356,950, 1973) for the same amount of waste burned.

A crosscheck for the value of heat recovered in 1975 is the actual revenue from the annual energy balance report. The value of coal and natural gas is taken from the records of the Electricity Works for the actual average price for these fuels in 1975. The price paid for coal was 18.36 DM per Gcal ($1.88, 1975, per 10^6 Btu) while that actually paid for natural gas was 17.51 DM per Gcal ($1.79 per 10^6 Btu). Only 204,928 metric tons (225,741 short tons) of waste were actually burned at the North plant in 1975, while at the South plant, 243,500 metric tons (268,285 short tons) were burned. The revenues from heat recovery as forecast and revised were thus:

CATEGORY	(In thousand DM) NORTH Original	Revised	SOUTH Original	Revised
Value of Heat Recovered	4,786	4,891	3,623	4,818

Forecast revenues from scrap sales were based on a recovery rate of 1.2% by weight and a sale price of 15 DM per metric ton ($5.10, 1973, per short ton). Actual recovery was 1.6%, or 3,333 metric tons (3,672 short tons) in the North and 2%, or 4,789 metric tons (5,275 short tons) in the South. The average sale price was 20 DM per metric ton ($7.50, 1974, per short ton). The forecast and revised revenues from scrap recovery were therefore:

CATEGORY	(In thousand DM) NORTH Original	Revised	SOUTH Original	Revised
Scrap Metal Sales	42.75	66.66	42.75	95.78

These accounts do not include the cost of the ash hauling operation (trucks and drivers) or any charge to the Electricity Works for dumping the ash at the municipal dump.

The total cost and revenue statement, with the exceptions noted, may be summed up as follows:

CATEGORY	(In thousand DM) NORTH Original	Revised	SOUTH Original	Revised
Total Costs	9,281	15,893	8,060	14,883
Total Revenues	4,829	4,957	3,666	4,914
Difference	4,452	10,936	4,394	9,969
Metric Tons Burned	237,500	204,928	237,500	243,550
Short Tons Burned	261,621	225,741	261,621	268,285
Cost Difference per metric ton	18.75 DM	53.37 DM	18.50 DM	40.93 DM

Thus according to the 1973 report, on the basis of 237,500 metric tons (261,621 short tons) burned in each plant, a rough cost average of 18.80 DM per metric ton ($6.39 per short ton) would remain to be covered by the city after heat and scrap metal were recovered. The real costs for 1975 (still understated, perhaps, in some of the personnel areas) however, even after allowing for higher compensation for heat recovered and scrap metal, showed deficits of 53 DM per metric ton ($14.56 per short ton) at the North plants and 41 DM per metric ton ($15.13 per short ton) at the South plants.

These extra costs were absorbed without difficulty by the Electricity Works because electricity rates are such that this section of the City Works has a large operating surplus each year. This surplus helps the city compensate for other services, such as traffic and transportation, which are always in the red. The higher cost of waste treatment seems to have been absorbed by the higher revenues from the sale of electricity and heat, but the cost differential credited to the Electricity Works as per the 1973 estimate report appears far from covering the real gap.

System Management and Operational Control

The North and South waste-burning incinerators are managed as integral parts of the electrical generation and district heating programs in Munich. Each waste-burning plant is under the control of a local director and staff whose primary function is to insure the delivery of electricity and heat. All of the plants are under the overall supervision of the Electricity Works general staff.

The suppliers of equipment to the Munich Electrical Works were required to provide detailed guarantees of performance and equipment life not only for the waste-burning grates, but also for the other special equipment. In the case of Martin grates, these guarantees are impressive multipage documents listing performance standards that had to be reached before plant acceptance. In some cases, the manufacturers were also required to guarantee the life of their equipment for a certain length of time or number of operating periods.

Plans for the Future

The Munich City Authorities and the Electricity Works directors are now planning to increase waste incineration capacity. A new federal law on solid waste management was passed in 1972, and amended in 1974. This law required each Lande to be responsible for its overall solid waste planning and to make final decisions in that area.

The Lande of Bavaria subsequently passed legislation in June 1973, and set out to comply with the federal law by reviewing existing solid waste treatment practice and planning for the future. While no definite plans have yet been adopted, it is likely that the Munich region will be responsible for solid waste treatment for a number of neighboring communities. This, plus the expected growth of the city itself, and the likely increase in per capita solid waste generation all suggest that present capacity is not sufficient.

And yet, considering the installed capacity of 3 units rated at 40 metric tons (44 short tons) per hour and 2 others at 25 metric tons (27.5 short tons) per hour, it may appear surprising that more capacity is really required. The present capacity operating 24 hours a day and 365 days per year at full load could treat 1,489,200 metric tons (1,640,449 short tons) per year. Since the expected requirement by 1980, even from a greater area, will probably not exceed 1,120,000 metric tons (1,233,751 short tons), why shouldn't this suffice?

The explanation is that at present only 37.3% of the capacity (based on waste having heat value of 1,500 kcal/kg or 2,700 Btu/lb) is used. The remaining 62.7% is accounted for by a series of constraints on the system. The first is an operating rate reduced by 26% to diminish corrosion and wear.

The second is a reduction by about 26% due to electrical load requirements which are such that the waste-fired plant cannot run at full load all the time. As a result, there is a further loss of capacity during start-up, shut-down and load adjustment periods.

The third is a reduction of 9.7% due to annual repair and maintenance. The last constraint is a 1% loss of capacity because the bunker at the South plant is too small.

The maximum that could be expected from the system, if the demand for electricity were perfectly constant and if there were no start-up, shut-down or load adjustment losses, would be still only 63.3%.

The annual heat requirements for electricity and for district heating vary considerably from month to month as shown in Figure 5.13. The contribution of waste to meeting this requirement is apparently the most constant of all the fuels used.

The daily electrical demand by hour is shown in Figure 5.14. There is an extreme peak load at about 8:30 AM and low trough at 4-5 AM. But, the city also faces some bizarre institutional constraints. During the early morning hours when city demand is low, additional electrical energy can be produced and delivered to the external power grid, but the city is not paid for this. The city has the right to generate its own power, but the surrounding area of Bavaria is the exclusive domain of the Bavarian Power Works. The relations are such that the city must buy power from the Bavarian Power Works when it needs it, but cannot sell it when it has excess capacity.

Because of the highly fluctuating daily electrical demand and the resulting loss in waste-burning capacity, and because of other capacity losses due to a need to diminish corrosion, the city authorities have decided to build no more combined waste and fossil fuel power plants.

FIGURE 5.13: ENERGY REQUIREMENTS OF MUNICH BY MONTH IN 1975 (for heat and electricity)

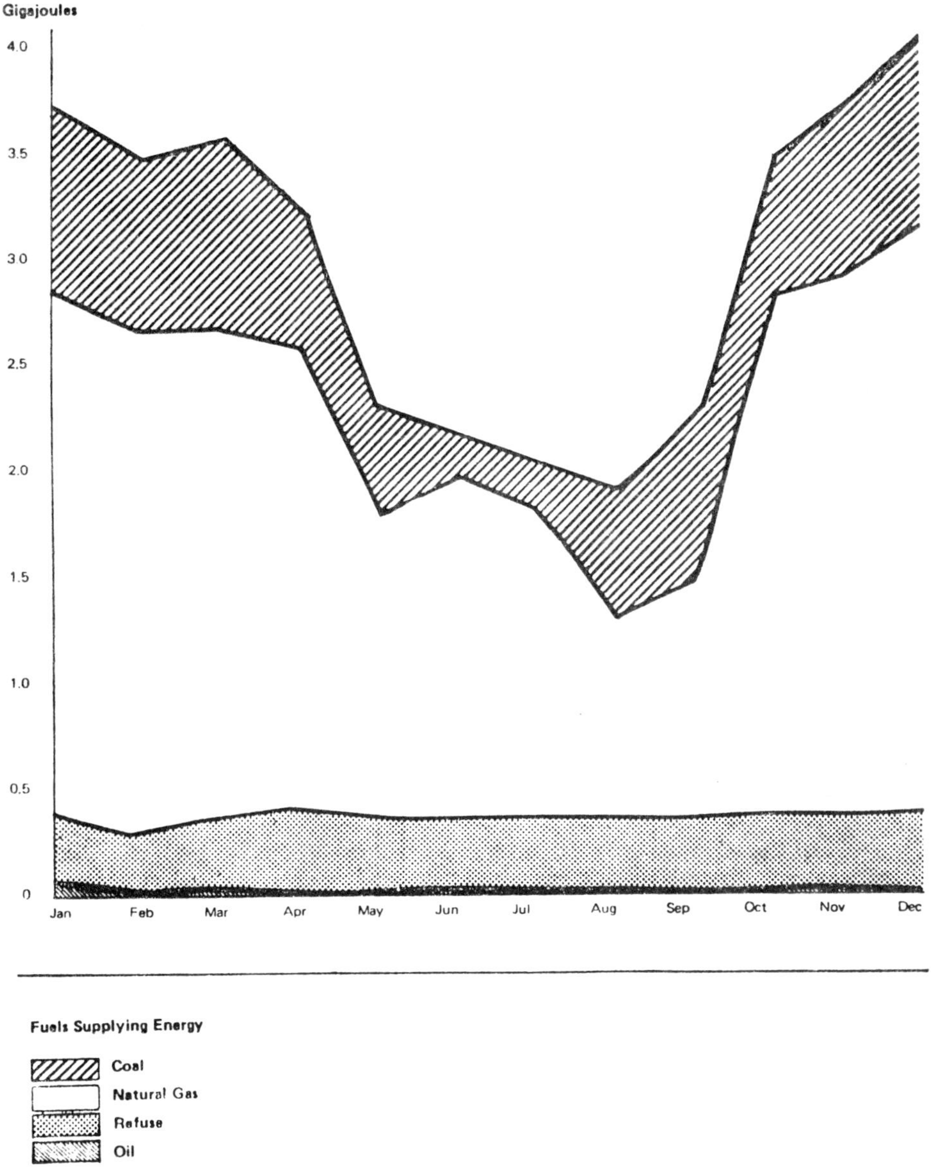

FIGURE 5.14: DAILY POWER REQUIREMENTS IN MUNICH BY HOUR OF THE DAY DECEMBER, 1973

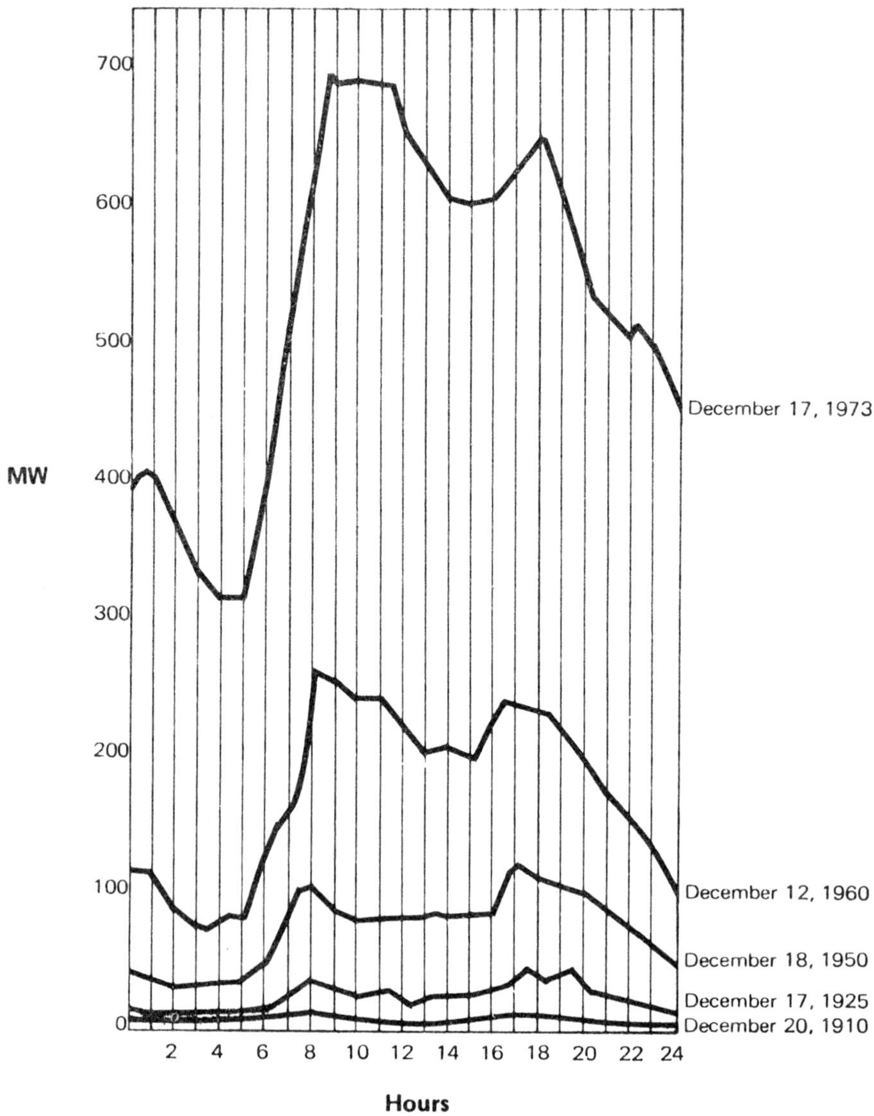

Munich has experimented with three different complicated designs for integrated waste and fossil fuel firing of Benson boilers. The city is now abandoning this solution, and moving instead toward all waste-fired units with a lower rate of energy recovery. The new plants will be designed so that some of the recovered heat will serve to dry sludge from the water treatment plants. They will incorporate burners to fire this sludge in a manner similar to the firing of pulverized coal on the North I and II units.

The new units (requests for proposals are now out) will have a capacity of 25 metric tons (27.5 short tons) per hour (like the original North I units). The boilers will not be Benson boilers and the steam will be at much lower pressure and temperature, although still superheated. Turbines will still be coupled for the generation of electricity, but they will be small, about 20-25 MW, and the steam will not be reheated.

CONCLUSIONS

It is difficult to draw simple conclusions about the various combined waste and fossil fuel electrical generation plants in Munich because there are many different perspectives from which comments need to be made. Cogeneration of heat and electrical energy is clearly feasible. The relative success of the North II unit suggests that the biggest problems may be institutional rather than technical.

Technical Conclusions

All three of the basic designs for the plants in Munich were original, first-ever conceptions for very large and very complicated plants. The city authorities, the Electricity Works planners, and the equipment suppliers showed an admirable willingness to experiment with radically new concepts and to bear the consequences which could result.

The first approach, used in the North units, has waste and coal fired in separate chambers but with the gases joining and a common heating of boiler elements. In these two plants there were several design problems which led to later difficulties. The coal chamber was too small and eventually there was corrosion in the second pass of the boiler unit. The waterwalls did not extend down to the grate in the waste portion of the furnace, resulting in excessive heat and subsequent ash fusion on the refractory walls. Martin considers this a design fault, but perhaps Babcock would disagree.

North IA and IB had serious problems with corrosion in the boiler tubes, particularly in the waste portion of the furnace. There were several attempts to diminish these corrosion problems, but they were never fully overcome. Both Martin and the Electrical Works personnel concede that this design was interesting, but not one which they would repeat.

Munich North II, with combined waste and coal firing of a common firebox seems to have been a much better solution. Corrosion problems have been less severe than at North I, and, in general, this plant has proven reliable and efficient. The basic vulnerability is that the waste portion is intimately linked to the rest of the plant, hence, if for any reason there is a difficulty with the waste portion, the whole plant, including the turbo-generator, is in jeopardy. Thus the risk of the unavailability of the whole plant is greater than if the waste unit were not an integral part of the plant.

Martin claims that the North II design is probably the best from several points of view including cost, reliability, and flexibility in the firing rate to suit power fluctuations. The company claims that North II is still a very efficient form of heat recovery from waste.

The City Electrical Works personnel, including the recently retired technical manager of all Munich power plants, do not share this enthusiasm. They concede the merits of this plant but feel that there is too much risk in having the waste plant compromise the generation of electricity. This argument cannot be settled easily, as it is influenced not only by operating history but also by the psychology of those responsible for delivering electric power when it is required.

The Munich South units represent a departure from the combined firing philosophy of the North units. In the South, the heat from waste is absorbed in an economizer unit for a subsequent main boiler which delivers the high pressure and temperature steam to the turbo-generator. This design incorporates a complete redundancy of the economizer capacity of the waste-fired boiler.

There are some difficulties in this inclusion of an external economizer in a Benson boiler arrangement, however. They involve a series of complications in running the different parts of the plant at different rates of output, in starting up the plant, in varying the load levels, and in operating the waste portion of the plant alone.

The plant was designed to enable the waste furnace to deliver up to 20% of total heat required at full plant output. However, when the plant is operating at less than full power, the waste portion has to reduce its output rate which means that waste incineration is not a function of waste availability, but follows the load of the power plant. There have been attempts to operate the waste portion of the plant when the fossil fuel portion is not on line, but this has proved complicated and possibly dangerous.

Martin claims that this design is good and that it has proven itself in service. They complain about the refusal of the boiler maker to bring the waterwalls down to the level of the grates, but feel that this design could be applied elsewhere. The Electrical Works authorities also feel that the design is basically sound, but suggest that it may not be suitable for Munich. A better load following capability and better start-up and shut-down possibilities would be needed as well as the capacity to burn waste by itself.

Comparing the overall efficiencies of the three types of plants in terms of conversion of heat from waste into electricity, the energy balance report for 1975 showed that North II was the most productive, followed by South IV. Each of these plants represents the most evolved form of its design group.

Economic Conclusions

From the information available, it is difficult to make conclusions about the economics of the various plants in Munich. The North I units would not be proposed for further applications but North II and the South plants seem attractive in the light of the costs published by the city. It is apparent, however, that these are not the real costs and that some of the actual cost is hidden and subsidized by the buyers of electricity.

Rough revision of the costs show that they are probably 2-3 times higher per ton than the published figures. There are still many arbitrary attributions of direct costs (which man works on which part of the combined plant), arbitrary attributions of general overheads (how much the waste portion of the plant should contribute to general overhead), and assumed fictitious tax charges which could be paid, but are not, because the waste turnover and capital charges are buried in the larger total Electricity Works accounts.

There is no doubt that the revised cost estimates in the report will be challenged, and probably should be. However, they rely upon some of the internal accounting data of the city. It has been suggested that the political sensitivity of charges for waste treatment is greater than that of minor variations in the electricity rates. This is perfectly understandable. It appears that waste incineration is much more expensive than is generally believed and that electricity sales subsidize a major part of the cost. Only the true operating costs and revenues could indicate whether building similar plants in other cities would be desirable. Obtaining real and unambiguous costs would probably be impossible. Approximate costs can be estimated reasonably well.

Comments on Future City Plans

No general technical or economic conclusions can be drawn from the decision of the Electricity Works and city administrators to build no more combined plants, but only all-waste-burning plants with lower grade steam and small electrical generators. This decision seems dictated more by the peculiar electricity requirements of the city than by any conclusive inadequacy of combined fired units. The authorities claim that they will not build any more large power stations at all, and this seems to be the principle reason for abandoning the combined plant philosophy. Even in the new plants, however, there is an interest in technological experimentation, as these plants will recycle some of their own recovered energy by using heat to dry sludge which they will then burn to generate more heat.

There seem to have been no complaints about the performance of the major system components including the Martin grate, waste feeding systems, ash quenchers, the Lurgi electrostatic precipitators, or the AEG or Siemens turbogenerators.

REFERENCES

Literature

The following documents were found useful as sources of information about the activities of the Munich incinerators:

(1) *Resource Recovery Through Incineration,* and a series of other documents prepared in English and in German describing the activities of the Josef Martin Feuerungsbau GmbH Company and the plants that they have helped build.

(2) Brochures prepared by the Stadtwerke München Elektrizitätswerke describing the details of their North and South combined fired refuse plants, and others describing the general organization and operations of the Werke.

(3) *Die Konzeption des Heizkraftwerkes Sud München und Einige Betriebserfahrungen,* by Dipl.-Ing. Franz, an article describing in detail the Munich South plant written by the new director of planning.

(4) *Das Müllverbrennungskraftwerk Nord der Stadtwerke München-Elektrizitätswerke,* by Dipl.-Ing. Herman Seifert, from "Technik + München," No. 1, 1965.

(5) *Sieben Jahre Müllverbrennung in München,* by Eugen Stelze, from Stadtreinigung und Fuhrpark.

(6) *Die Müllverbrennungsanlagen der Stadtwerke München,* by Karl-Heinz Luscher, from Elektrizitatswirtschaft, Vol. 67, 1968 No. 18.

(7) *Müllkesser Heizen Münchens Neue Entlastungsstadt,* by Oberbaurat Burkhard Franz, from Energiewirtschaftliche Tagesfragen Vol. 18 (1968) No. 2.

(8) *Müllkraftwerk München,* by M. Andritzky, from Brennstoffe-Warme-Kraft, Vol. 14 (1962) No. 5 May 1962.

(9) *Müllverbrennung und Wärmeverwertung,* by F. Maikranz, speech given by Mr. Maikranz at a Refuse Incineration Conference in Hamburg on 9 February, 1968.

(10) *Erfahrungen Mit Der Verfeuerung Von Müll in Einem Hochdruck-Dampfkraftwerk,* by Dr. Herbert Bachl and Franz Maikranz, from Energie, Vol. 17, No. 8, August 1965.

(11) *Massnahmen Zur Vermeidung Von Rohrabzehrungen in Müllkesseln,* by K.H. Luscher, from a conference speech at Incineration of Refuse 1972, 18 April, 1972 in Essen, Germany.

(12) *Kostenprobleme Bei Müllverbrennungsanlagen,* by A. Presuhn, from Brennstoffe-Warme-Kraft, Vol. 19 (1967) No. 10.

(13) *Ermittlung der Kosten der Müllverbrennung in Den Kraftwerken Nord und Sud der Stadtwerke-Elektrizitätswerke,* fur die Zeit vom 1.1.1973-31.12-1975.

Persons Contacted

Mr. Franz,	Newly appointed head of Planning for the Heat Section of the Electricity Works, Munich.
Mr. Albat,	Newly appointed head of Operations and Maintenance of all of the Thermal Power Plants in Munich.
Mr. Schroeder,	Head of the Economic Group of the Heat Group of the Electricity Works.
Mr. Kraus,	Director of the Munich North Plant.
Mr. Stache,	Statistical Department of the Electricity Works.
Mr. Walter Martin,	Director of Josef Martin Feuerungsbau GmbH, major suppliers of equipment to the three plants.
Mr. Eugen Stelze,	Former director of Operations and Maintenance of all of the thermal power plants in Munich, served as a consultant to Consultex for this study.
Mr. Dieter Kreuch,	Engineer of Josef Martin Feuerungsbau GmbH, helped with arranging visits and with furnishing information about the plants.

Case Study of Brive, France

Material for this chapter has been based upon a report prepared by Resource Planning Associates, Inc. (RPA), Washington, D.C. (CONS/2103-5).

WASTE TREATMENT COMPLEX OF BRIVE, FRANCE

Brive is a city of 55,000 inhabitants located in the Department of Corrèze, a low-population-density area in the south of France. It is the largest municipality in the department, but Tulles, a smaller agricultural town, is the administrative center. Brive lies in the intersection of two main transportation (i.e., water, highway, railroad) axes, one north-south and the other east-west, which explains its development, after World War II, from a rural town to a center of agribusiness and industry.

Brive has an area of 42 km^2 (18.5 sq mi), with an average density of 1,150 persons/km^2 (2,978 persons/sq mi). Except for the massed housing in the city center, the city is composed mainly of individual houses. An industrial area is located in the southwest section of the city, and there are small businesses in the southeast. The main industries are food processing, heavy manufacturing, and electronics.

Household waste is collected three times a week from about 100 communities (160,000 inhabitants) in the Brive area. A public organization, the Syndicat Intercommunal de Ramassage et de Traitment des Ordures Ménagères (SIRTOM), collects and treats the solid waste of 80 communities (120,000 inhabitants) including Brive and surrounding towns. SIROM, a public waste collection organization in Tulle, and SICTOM, a similar operation in Sarlat, collect waste from 20 communities (40,000 inhabitants) and transport it to transfer stations in Tulle and Sarlat, about 40 km (25 mi) from Brive. At those points, SIRTOM picks up the waste and transports it to Brive. This arrangement was launched in October 1974. It is an unusual experiment for France, and has not been without political and administrative problems. However, it is expected that other communities will join this scheme and that new transfer stations will be created.

SIRTOM employs 56 workers and uses 18 packer trucks to collect the waste from Brive and the surrounding area. The area is divided into 18 districts, each served by a packer truck and a 3-person crew. In Brive, plastic bags are used for collection. All household waste is burned in the incinerator. Some special waste, such as bulky waste, tires, grass, and leaves, is dumped in a noncontrolled landfill.

Waste collection and transportation are paid for by local taxes. Each community contributes to the budgets of the collection organizations according to the number of hours of collection and the tonnage of waste transported. The cost of waste collection for Brive was FFr 1,381,600 (FFr 88/hr × 15,700 hr) ($289,000) in 1976 while the cost of transportation was FFr 429,000 (FFr 33 per metric ton × 13,000 metric tons) ($90,000). The cost of the plastic collection bags in 1976 was FFr 450,000 ($94,000). Thus, the total cost of waste collection and transportation in Brive was FFr 41 ($8.60) per inhabitant, or FFr 174 per metric ton ($33 per short ton) in 1976.

The collected waste is transported to the waste treatment complex in Brive. To gain a clear understanding of the nature and operation of this complex, the remainder of this case study explains:

- How the current waste treatment system was developed
- The design of that system
- The way in which the system operates
- System costs

DEVELOPMENT OF CURRENT WASTE TREATMENT SYSTEM

In the early 1960s, all of Brive's waste was dumped in a noncontrolled landfill. But because of complaints from neighbors and the frequent fires, the municipality decided in 1968 to find a new means of disposing of its solid waste. Controlled landfill was the least expensive solution, but no suitable location could be found (1). And composting was rejected because of the fluctuation of prices resulting from the small number of potential buyers (2).

At the same time, Brive also needed to build a sludge treatment facility. The possibility of a joint facility for solid waste incineration and sludge treatment was suggested by M. Val, the director of Brive's technical services, when he learned that such a plant was being built in the French city of Dieppe in 1969. M. Val, a young engineer who had visited a number of new waste treatment systems in France and other countries (including the United States), studied the technical and economic aspects of Dieppe's system and proposed certain changes. In particular, he felt that the incinerator should burn fresh instead of digested sludge, as at Dieppe. (Digested sludge is a mixture of sewage solids and water that has been decomposed by anaerobic bacteria.)

Preliminary studies were conducted to locate a suitable site for the solid waste and sludge treatment complex, and the municipality voted in favor of this combined system. Because this combination of solid waste incineration and sewage treatment was new, M. Val's persuasive arguments in favor of a joint facility were a decisive factor in making this solution acceptable at all levels. The fact that the mayor of Brive, M. Charbonel, was also the Minister of Industry in the French national government was helpful.

Preliminary studies of the solid waste incineration unit and sewage treatment unit were carried out in 1968 and early 1969 by M. Val and the technical services unit. The results of these studies were the basis for the bid invitation issued at the end of 1969. This bid invitation specified a joint facility for solid waste incineration and sludge treatment, using energy (heat) recovered from incineration to dry the sewage sludge which would subsequently be incinerated with the waste (3). It also specified that the contract to be awarded would include not only construction but also operation of the incineration and sludge treatment facility. The operation phase was included to obtain a better estimate of the overall cost of waste treatments and to take advantage of a tax benefit—specifically, to recover the value added tax (17.6% of the capital costs)—from operating the facility through a concession.

Bidding was open to companies selected by prequalification. Five groups of companies were selected in June 1970. Six other companies that had qualified for only the incinerator or only the sewage treatment plant were asked to form groups and present joint bids (4).

By December 1970, eight groups of companies had been judged qualified (see Table 6.1). The municipality urged two of these firms, Société de Construction d'Usines d'Incinération d'Ordures Ménagères (INOR), the French representative of Von Roll of Zurich, and Omnium d'Assainissement (ODA), one of the major French specialists in water treatment, to present a joint bid, which they did in June 1971.

After eliminating several competitors for technical reasons, city officials had to decide between INOR/ODA and Degrèmont/SFRIR. Although the INOR/ODA bid was a little higher than that of Degrèmont/SFRIR, the equipment offered was considered more reliable. This was especially true of the vertical press filter and the vertical sludge dryers. Moreover, INOR had built the incinerator for the solid waste/sludge treatment complex at Dieppe. Consequently in July 1971, the contract for construction and operation was awarded to INOR for the incineration unit and ODA for the sewage treatment unit. Construction of the incinerator began in November 1971, while construction of the sludge treatment unit began in May 1974. The incinerator began operations in September 1973, and the sludge treatment unit, in February 1975.

TABLE 6.1: QUALIFIED GROUPS OF COMPANIES

Group No.	Incinerator Contractor	Waste-Water Treatment Contractor
1	INOR (representative of Von Roll)	INOR
2	Tunzini (representative of VKW)	Degremont
3	SFRIR	Degremont
4	CNIM (representative of Martin)	CIE
5	CEC	Omnium d'Assainissement
6	SOFICO	Omnium d'Assainissement
7	Stein Industries	Froment Clavier
8	Luchaire	Luchaire

DESIGN OF THE WASTE TREATMENT SYSTEM

The solid waste/sewage sludge treatment complex, an attractive design by a local architect, is situated on a 2.8 hectare (7 acre) tract in the western part of Brive. It is about 4.8 km (3 mi) from the town center, at some distance from any residential area.

The main building, which houses the incineration and sludge drying operations, has an area of 500 m^2 (5,380 sq ft) and a volume of 12,000 m^3 (424,000 cu ft). It is roughly L-shaped, with the operations and maintenance sections located in the middle and the filtration operation located in the inner angle of the L. A separate building contains a truck-weighing scale, a laboratory, heating equipment for the main plant, and an administrative office.

The waste treatment complex is composed of a waste incineration unit, an energy recovery operation, a sewage treatment unit, a sludge drying unit, and a multicyclone to control particulate emissions (see Table 6.2).

The incineration unit is equipped with two Von Roll furnaces, each capable of processing 3.5 metric tons (3.9 short tons) of waste per hour. [The originial contract had called for two furnaces, each with a capacity of processing 3 metric tons (3.3 short tons) per hour. However, a cannery producing 40 metric tons (44 short tons) of solid waste and sewage per day opened in the region. Consequently, the hourly capacities of the two furnaces were increased.] The two furnaces were designed to operate 7,000 hours per year, which would permit the incineration of 49,000 metric tons (54,000 short tons) of waste annually. In reality, the estimated quantity of waste treated in 1976 is 37,000 metric tons (40,750 short tons), or 75% of maximum capacity.

Energy is recovered by a Socomas fire-tube boiler on top of each furnace. The

TABLE 6.2: TECHNICAL DATA SHEET

Dates	
Start of construction	November 1971
First incineration of solid waste	September 1973
First incineration of sludge	February 1975

	Metric System	English Equivalent
Plant dimensions		
Area of site	2.8 hectares	6.9 acres
Area of building	500 m^2	5,380 sq ft
Height of furnace room	26 m	85 ft
Height of smokestack	38 m	125 ft
Diameter of smokestack	1.35 m	4.4 ft
Volume of waste bunker	750 m^3	981 cu yd
Volume of crane bucket	1.5 m^3	2 cu yd
Capacity of crane bucket	4 metric tons	4.4 short tons
Capacity of scale	30 metric tons	33 short tons
Incinerator		
Number of furnaces	2	
Grate area, per furnace	16 m^2	172 sq ft
Thickness of refractory	50 cm	1.6 ft
Rated capacity, per furnace	3.5 metric tons per hour	3.9 short tons per hour
Volume of primary combustion air, per furnace	20,000 m^3/hr	706,200 cu ft/hr
Volume of combustion gas, per furnace	25,000 m^3/hr	882,750 cu ft/hr
Temperature of combustion gas	800°C	1472°F
Temperature of flue gas, at exit from boiler	350°C	662°F
Capacity of induced draft fan	25,000 m^3/hr	882,750 cu ft/hr
Efficiency of multicyclones	60%-70%	
Daily capacity, for plant	150 metric tons	165 short tons
Heat recovery system		
Number of boilers	2	
Steam production, per boiler	9.7 metric tons per hour	10.7 short tons per hour
Rated pressure	16 bar	15.8 atm
Working pressure	10,5 bar	10.4 atm
Working temperature	193°C	379°F
Number of sludge dryers	3	
Capacity of sludge dryers (from 89% moisture content to 45% moisture content)	2.5 m^3/hr	3.3 cu yd/hr

In addition to INOR, the following companies participated in the construction of the incineration unit:

Construction	Roussel–Brive
Structural steel work	Barbarin–Ruffec
Lighting	Sobel–Brive
Locksmithing	Sanz–Brive
Plumbing and heating	Francy–Brive
Facade	Voisin
Steelwork for machinery	Barbarin–Ruffec
Overhead crane	Reel–Lyon
Ventilation and air filtration	Prat–Paris
Ash removal	Sanz–Brive
Electrical work	CGEE–Bordeaux
Condensors	Creusot Loire –Paris
Boiler	SOCOMAS–Hautmont

SOURCE: INOR

rated capacity of each boiler is 9.7 metric tons (10.7 short tons) of saturated steam at 16 bars (15.8 atm) per hour. However, steam is actually produced at 10.5 bars (10.4 atm).

The steam is condensed in the sludge dryers, in the heat exchangers for incoming combustion air, or in air-cooled condensers. At present, from 4 to 4.5 metric tons (4.4 to 5 short tons) of steam per hour are not used and simply condensed. Condensate returns to the boiler, so that the boilers, the sludge dryers, the air preheaters, and the condensers form closed systems. The feedwater is drawn through a treatment unit which has a capacity of 2.5 m^3 (660 gal) per hour.

The energy recovered during incineration is used to dry sludge from the sewage treatment unit in three vertical Luwa dryers. Each dryer has a rated capacity of drying 2.5 m^3 (3.3 cu yd) of sludge per hour from 89% moisture content to between 40% and 50% moisture content. This dewatered sludge can be added to the regular solid waste for incineration.

The exhaust gases pass from the incinerator through multicyclones located in the main building, to remove particulate emissions before escaping into the air through a smokestack 38 meters (125 ft) high. These multicyclones, which have an efficiency of 60 to 70%, keep particulate emissions below 600 mg/m^3. The CO content is less than 0.5%.

OPERATION OF THE WASTE TREATMENT SYSTEM

Brive's waste treatment system is a relatively complex operation. To ensure complete understanding, we describe the system in detail, explain how it is managed, and summarize the results of evaluating its efficiency and reliability to date.

Description

The waste treatment has four distinct operations: incineration of solid waste (including dried sludge), treatment of wastewater, heat recovery and sludge drying, and emissions control. The interrelationship of the first three can be expressed as follows:

1 metric ton of waste with a lower heating value of 1,500 kcal/kg —(thermal efficiency 65%)→ 1.8 metric tons of saturated steam at 10.5 bars

↓

1 m^3 of sludge, dried from a moisture content of 89% to a moisture content of 45% → 0.18 metric ton of dried sludge with a moisture content of 45% and a lower heating value of 770 kcal/kg

With the energy generated from incinerating 31,000 metric tons (34,000 short tons) of waste in 1975, it would have been possible to dry approximately 31,000 m^3 (8,190,200 gal) of sludge (at 89% moisture content). However, because sewage comes only from Brive, only about 10,000 m^3 (2,642,000 gal) of sludge was dried. This operation produced 1,900 metric tons (2,090 short tons) of dried sludge for subsequent incineration (5). In 1976, approximately 37,000 metric tons (40,750 short tons) of waste were incinerated. With the energy generated, it would have been possible to dry approximately 37,000 m^3 (9,775,000 gal) of sludge with a moisture content of 89%. However, only an estimated 16,000 m^3 (4,227,200 gal) of sludge were dried, producing 3,000 metric tons (3,300 short tons) of dried sludge.

Each of the four major operations is detailed in the remainder of this section.

Incineration: Solid waste is brought by truck to the incinerator (see Figure 6.1) where it is weighed and then dumped into a waste pit with a volume of 750 m^3 (981 cu yd). This capacity allows up to 3 days' collected waste to be stored at any given time. There is no shredder.

A 4 metric ton (4.4 short ton) crane with a 1.5 m^3 (1.96 cu yd) bucket picks up the waste from the waste pit and puts it into the furnaces' water-cooled hoppers (1.80 m × 1.20 m or 5.9 ft × 3.9 ft), which have cast iron doors and feeders (6). These hoppers also receive the dewatered sludge from the dryers. The furnace has a steel frame and plates, and is insulated with refractory bricks to a thickness of 50 cm (20 in).

From the hoppers, the waste and dried sludge drop onto the distribution grate; are moved to the combustion grate, through the use of two sections of fixed, hydraulically-driven chrome steel bars; and transported, via two more sections of the fixed and movable chrome steel bars, to the burn out grate. There is a 1.2 m (3.9 ft) step between the combustion grate and the burn out grate to ensure better combustion of the waste. The total area of the grates is 16 m^2 (172 sq ft).

At the combustion grate, air, drawn from the waste pit and heated by a heat exchanger to a temperature of 150°C (302°F), is introduced under the grate and forced through the grate by a fan with a capacity of 20,000 m^3/hr (706,000 cu ft/hr). An oil-fired burner (150 kg or 330 lb per hour) is used to initiate combustion. The thermal limit of each furnace is 7.3 Gcal (28.9×10^6 Btu) per hour, which, in turn, imposes the 3.5 metric ton (3.9 short ton) per hour limit on capacity. Combustion temperature is 900°C (1,620°F).

A fire-tube boiler is located on top of each combustion chamber so that the furnace and the boiler form an integrated system. Secondary air is introduced before the boiler to ensure complete burning of the gases before they are exhausted through the stack.

Combustion residues fall through four hoppers into a water-filled trough

FIGURE 6.1: SECTION OF THE INCINERATOR AND SLUDGE DRYING EQUIPMENT

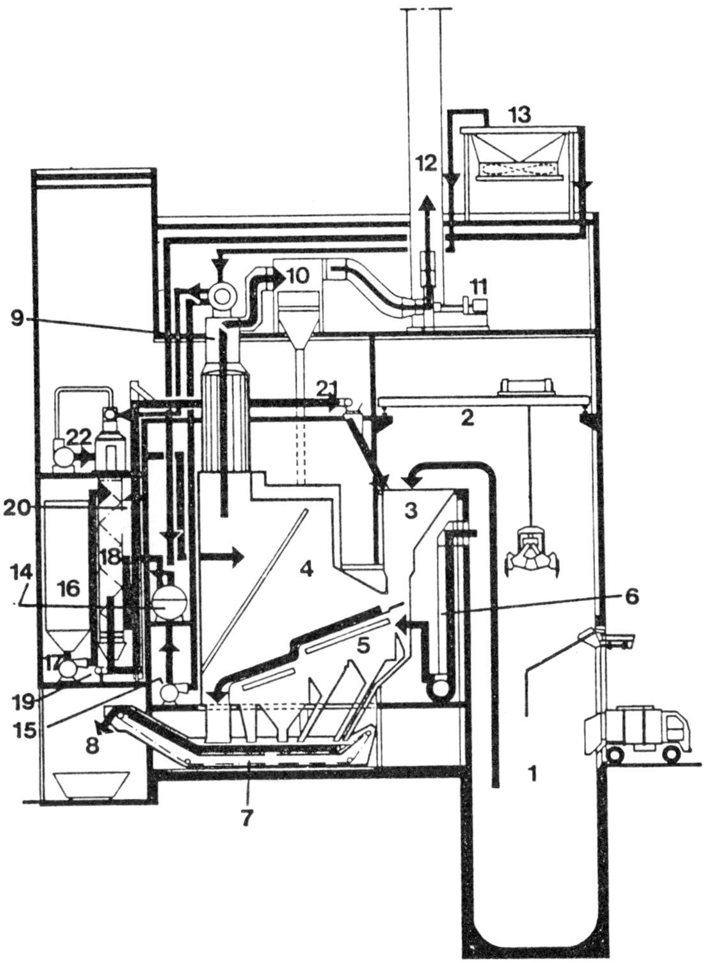

1. Waste pit	12. Stack
2. Crane	13. Condensor
3. Hopper	14. Condensate reservoir
4. Combustion chamber	15. Condensate pump
5. Grates	16. Sludge reservoir
6. Combustion air fan	17. Sludge pump
7. Ash trough	18. Sludge dryer
8. Ash discharge	19. Conveyor belt
9. Boiler	20. Hoist
10. Cyclone battery	21. Conveyor belt
11. Induced draft fan	22. Forced draft fan

SOURCE. INOR

which seals the furnace. They are carried by a conveyer chain to a storage bin. Eventually, they are trucked to the landfill.

Wastewater Treatment: Brive's combined sewage system handles residential and industrial sewage and storm drainage for the equivalent of a population of 160,000. The average levels of pollution are 208 mg/l of BOD_5 and 216 mg/l of suspended solids. The sewage, which contains 7.9 metric tons per day (8.7 short tons/day) of BOD_5 and 8.2 metric tons/day (9 short tons/day) of suspended solids, is transported through the sewers to the wastewater treatment unit of the waste treatment complex. The treatment unit has an average daily flow of 38,000 m^3/hr (10,039,600 gal/hr) and a maximum hourly flow of 3,625 m^3/hr (957,700 gal/hr).

Wastewater enters the treatment facility through a lift station with three screw pumps, each having a capacity of 1,400 m^3 (370,000 gal) per hour (see Figure 6.2). The waste is "screened" by two rotating screens, each 2 m (6.5 ft) in diameter. Grit is removed by a mechanical scraper with an area of 81 m^2 (872 sq ft) and a classifier. The wastewater continues to a primary sedimentation tank with a diameter of 50 m (164 ft) and a capacity of 6,850 m^3 (1,810,000 gal). After the wastewater is skimmed and is allowed to remain in the sedimentation tank for 2 to 3 hours, during which time it loses 30 to 35% of its BOD_5 and 60 to 65% of its suspended solids, it is moved to aerating tanks, which have total capacity of 4,200 m^3 (1,110,000 gal). Six ODA Neyrpic turbine aerators, each with a diameter of 2.8 m (9 ft) and driven by a 100-hp motor, activate the sludge in these tanks. The sludge then moves to a secondary sedimentation tank with a diameter of 50 m (164 ft) and a capacity of 7,850 m^3 (2,070,000 gal), but is recirculated into the aerating tanks, if necessary, to maintain the desired concentration of activated sludge regardless of the degree of pollution in the wastewater as it enters the treatment unit.

The effluent is piped into the nearby Corrèze River, while the sludge from both the primary and secondary sedimentation tanks goes to a thickener, having a diameter of 20 m (66 ft) and a capacity of 940 m^3 (248,000 gal). In the thickener, the moisture content of the sludge is lowered to approximately 94% (7). Some of the thickened sludge goes to a vertical press filter for chemical treatment by polyelectrolysis, which reduces its moisture content to approximately 80%. (However, the shortage of sludge, mentioned above, has led the plant operator to bypass filtration for some of the sludge.) Sludge from this process (80% moisture content) is combined with sludge that comes directly from the thickener (94% moisture content) to obtain a homogenous mixture with a moisture content of 88.6%. This mixture is pumped into the dryers. There has been no attempt to lower the moisture content of the sludge any further because the vertical thin film dryers require that the sludge remain relatively fluid at entry.

Heat Recovery and Sludge Drying: The thickened sludge from the wastewater treatment unit is pumped to tanks located at the tops of the three stainless steel vertical thin film sludge driers (each of which has a capacity of 2.5

FIGURE 6.2: SEWAGE TREATMENT

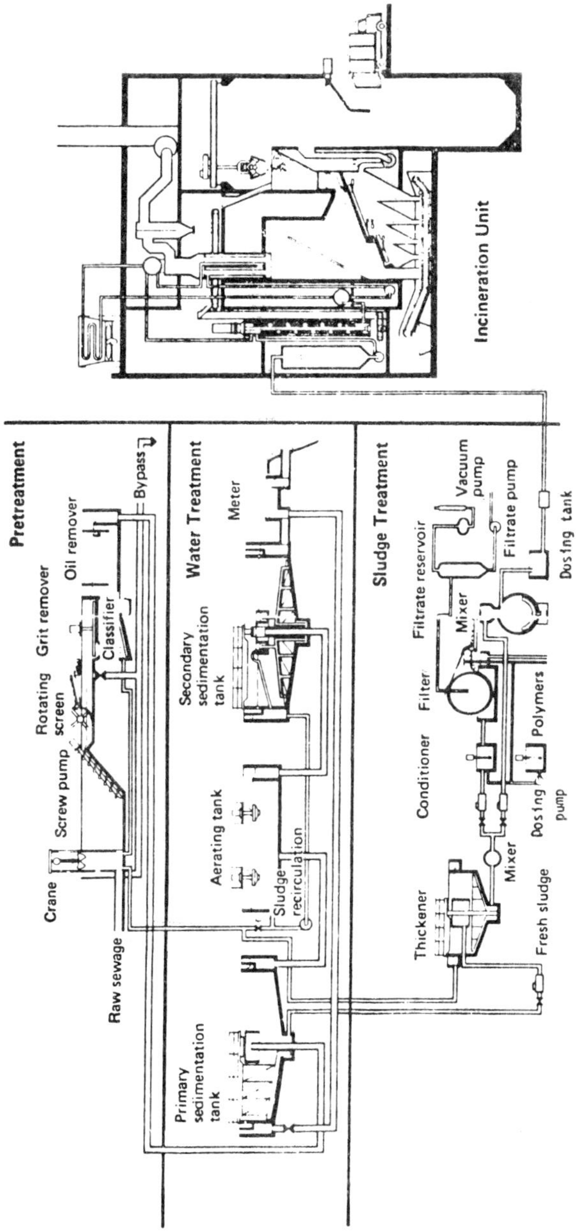

SOURCE: ODA

m^3/3.3 cu yd per hour of sludge) in the drying unit, which is located in the same building as the incinerator (see Figure 6.1). The sludge is then pumped into the top of the dryer cylinder. As it flows down through the cylinder, the sludge is spread over the inner surface by a rotor carrying self-adjusting vanes, revolving at a speed of 250 rpm (8). The cyclinder is surrounded by a steam jacket fed from the boilers of the waste-burning furnaces (see Figure 6.3). That is, the heat recovered during incineration enters the boilers, which have a heat exchange surface of 240 m^2 (2,582 sq ft) apiece, at 700°C (1,282°F). Each boiler produces 9.7 metric tons (10.7 short tons) per hour of steam at a pressure of 10.5 bars (10.4 atm). The steam is used to dry the sludge and the dryers thus serve as condensers for the boilers.

FIGURE 6.3: VERTICAL THIN FILM SLUDGE DRYER

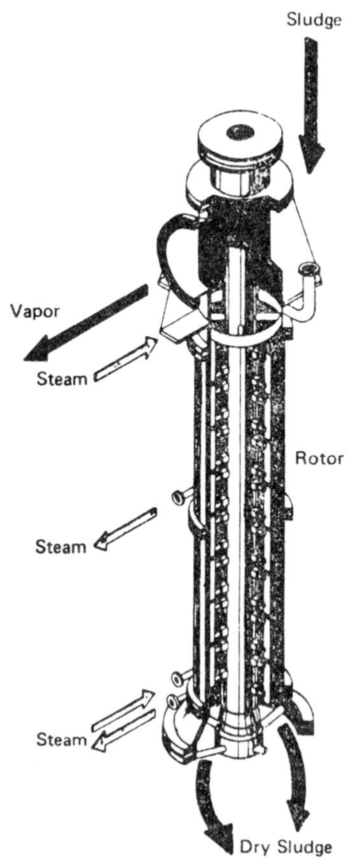

The vapors generated in the dryers travel upward and are extracted by fans through outlet tubes at the dryer heads. They pass into the incinerators as overfire air where they are deodorized at a temperature about 800°C (1,472°F).

The dewatered sludge contains only 40 to 50% moisture, and has a lower heating value of approximately 770 kcal/kg (1,390 But/lb). Thus, it can be incinerated along with the solid waste. To this end, dried sludge granules fall onto a conveyor belt under the dryers (see Figure 6.1), and are carried automatically to the hopper of one of the furnaces. In this operation, even distribution and constant feed are necessary for total combustion and satisfactory furnace operation.

Gaseous and Solid Emissions Control: Exhaust gases from incineration are produced at the rate of 25,000 m^3 (882,750 cu ft) per hour per furnace. The gas leaves the boiler at a temperature of 350°C (662°F) and is drawn through the multicyclone by a fan with an output of 28,000 m^3 (988,680 cu ft) per hour. The multicyclone reduces particulate emissions to an average of 350 mg/m^3 at 7% CO_2 which corresponds to an efficiency of 60 to 70% (9). These levels are in accord with the French legislation for plants of this size (*Circulaire du 6 juin 1972*). The gas escapes through a 38 meter (125 ft) smokestack having a diameter of 1.35 meters (4.4 ft). The smokestack is separated into two flues, corresponding to the two furnaces, to obtain a constant flow even when only one furnace is in operation. The CO content is guaranteed to be under 0.5%.

Particulate emissions were measured at the stack level at the inauguration of the plant (i.e., in October 1973). The tests, which were carried out by the Louis Prat Company (Paris), showed that the emissions contained 17% water, 4 to 6% CO_2, between 184 mg/m^3 amd 545 mg/m^3 of particulate, and had a pH of approximately 2 (see Table 6.3) (10).

TABLE 6.3: LEVEL OF PARTICULATE EMISSIONS

	Measurement Number	Operating Capacity of Furnace (metric tons*/hr)	Gas Temperature (°C)	Average Speed (m/s)	Gas Output (m^3/hr)	Particulate Level (mg/m^3)
Furnace No. 1	1	3.5	230	14.9	20.800	545
	2	3.5	230	14.8	20.600	225
	3	3.5	230	16.5	23.000	430
	4	3.5	230	16.4	22.900	184
	5	2.0	200	8.4	12.500	390
	6	2.0	200	8.8	13.100	383
	7	2.0	230	14.2	19.800	282
	8	2.0	230	9.2	13.150	500
Furnace No. 2	1	3.5	250	10.8	14.500	360
	2	3.5	250	10.8	14.500	260
	3	3.5	230	10.8	15.150	230
	4	3.5	250	12.3	16.500	270
	5	3.5	250	12.3	16.500	295
	6	2.0	200	8.6	12.800	450
	7	2.0	215	12.0	16.900	395
	8	2.0	200	6.8	10.150	495

SOURCE: INOR records.

*To convert metric tons to short tons, multiply by 1.1.

In addition to the exhaust gases, which contain some air pollution elements, incineration residue is produced in the waste treatment process. On average, incineration residue equals 37%, by weight (16% by volume), of the waste treated (11). The water content is about 50%. [The incinerator manufacturer guarantees an average of less than 0.5% putrescible materials and less than 5% unburned materials (12), but no measurement has been made to see if these limits are realized.] This residue has been used to build embankments, although the presence of steel cans limits this application. The recovery of ferrous materials from the ashes has also been considered, but it is economically infeasible in such a small operation.

Management

INOR manages the incinerator and sludge dryer and ODA operates the sewage treatment unit, in accordance with the 20-year concession in the original contract. However, INOR takes the lead in managing the complex, and the chief of operations is an INOR employee who lives at the site (see Figure 6.4). Brive exercises its overall control through its technical service unit, which is in direct contact with the chief of operations, who provides the city with monthly operational records and yearly technical reports (13).

Fifteen people are employed at the waste treatment complex: the chief of operations; 11 people employed by INOR at the incineration unit; and 3 people, employed by ODA, at the sewage treatment unit (see Table 6.4). (This dual employment structure has apparently led to problems of competency and coordination; as a result, in practice, each group works almost independently.) Of the 11 people working at the incineration unit, 6 operate the incinerator and sludge dryers, 2 are maintenance men, 1 is a janitor, 1 is a clerk, and 1 is a truck driver. The 6 incinerator operators are organized into three 2-man shifts, each working 8 hours a day. In each shift, one worker controls furnace operations and the other introduces the waste into the furnace. The clerk supervises the weighing of trucks and the truck driver hauls the incineration residue to the dump. INOR provides a replacement furnace operator during vacations.

At the beginning of operations, Von Roll and INOR engineers spent only about 3 months instructing the personnel in the proper operation and maintenance of the highly automated furnaces. According to the chief of operations, a worker can learn to operate the incinerator and sludge dryers in a few months. However, much more experience is necessary to deal with the operating and maintenance problems that arise.

Evaluation of System Efficiency and Reliability

The incinerator began operations at the end of 1973 and burned only waste during 1974. In 1973, 1,300 metric tons (1,430 short tons) of waste were incinerated. In 1974, the first full year of operations, 23,100 metric tons (25,440 short tons), including 20,800 metric tons (22,900 short tons) of household waste and 2,300 metric tons (2,530 short tons) of commercial and industrial

FIGURE 6.4: ORGANIZATION CHART OF THE PLANT WORK FORCE

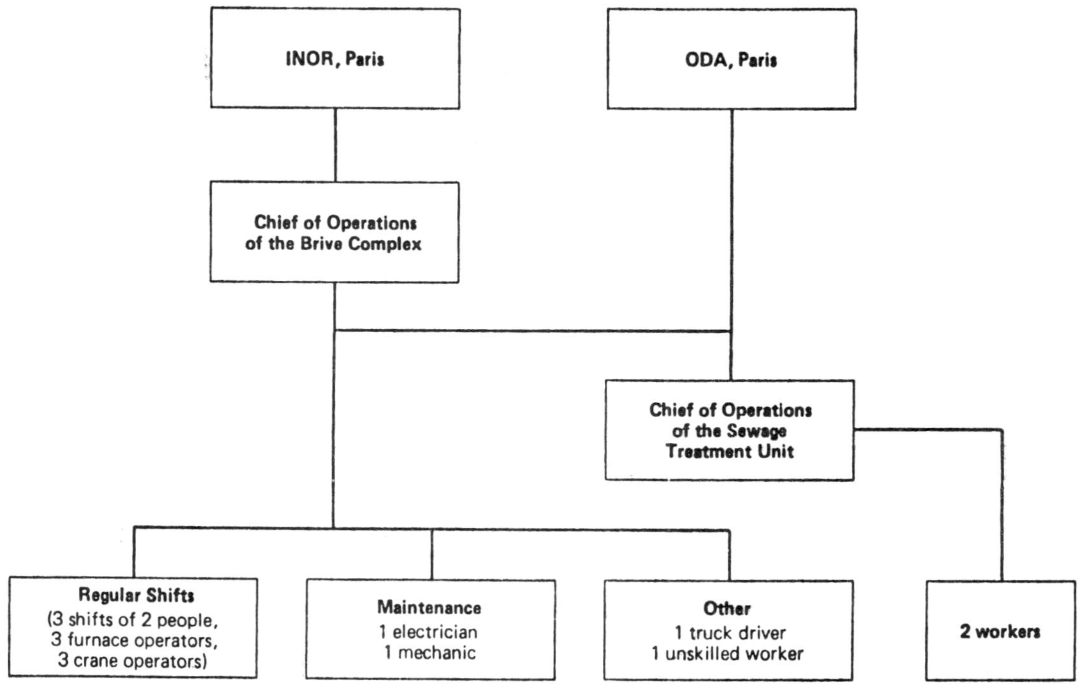

SOURCE: INOR

TABLE 6.4: MANPOWER SCHEDULE FOR THE INCINERATION PLANT IN 1976
(hours)

	Furnace Operation	Sludge Dryers Operation	Routine Maintenance	Nonroutine Maintenance	Administration	Others	Total
January	676	320	482	118	359	–	**1,955** (11 employees)
February	757	240	382	40	335	–	**1,754** (11 employees)
March	876	160	535	20	–	–	**1,591** (11 employees)
April	826	160	607	60	372	–	**2,025** (11 employees)
May	721	200	320	120	351	314	**2,026** (12 employees)
June	940	120	405	–	349	170	**1,984** (12 employees)
July	978	88	135	108	346	178	**1,833** (10 employees)
August	983	70	181	30	183	136	**1,583** (9 employees)
September	1,000	80	240	–	276	–	**1,596** (11 employees)
October	1,016	80	255	–	265	–	**1,620** (10 employees)
Total	**8,773**	**1,518**	**3,542**	**496**	**2,836**	**798**	**17,967**

waste brought in by private companies were burned. In 1975, a total of 29,100 metric tons (32,070 short tons), including 27,400 metric tons (30,200 short tons) of household waste collected by the waste collection organizations and 1,700 metric tons (1,870 short tons) of commercial and industrial waste was incinerated. (The monthly variations during that period are shown in Figure 6.5). During the first 10 months of 1976, 27,000 metric tons (29,740 short tons) of waste burned, including 25,600 metric tons (28,200 short tons) of household waste and 1,400 metric tons (1,540 short tons) of commercial and industrial waste (see Table 6.5).

FIGURE 6.5: MONTHLY VARIATION IN QUANTITY OF WASTE AND SLUDGE INCINERATED (1975)

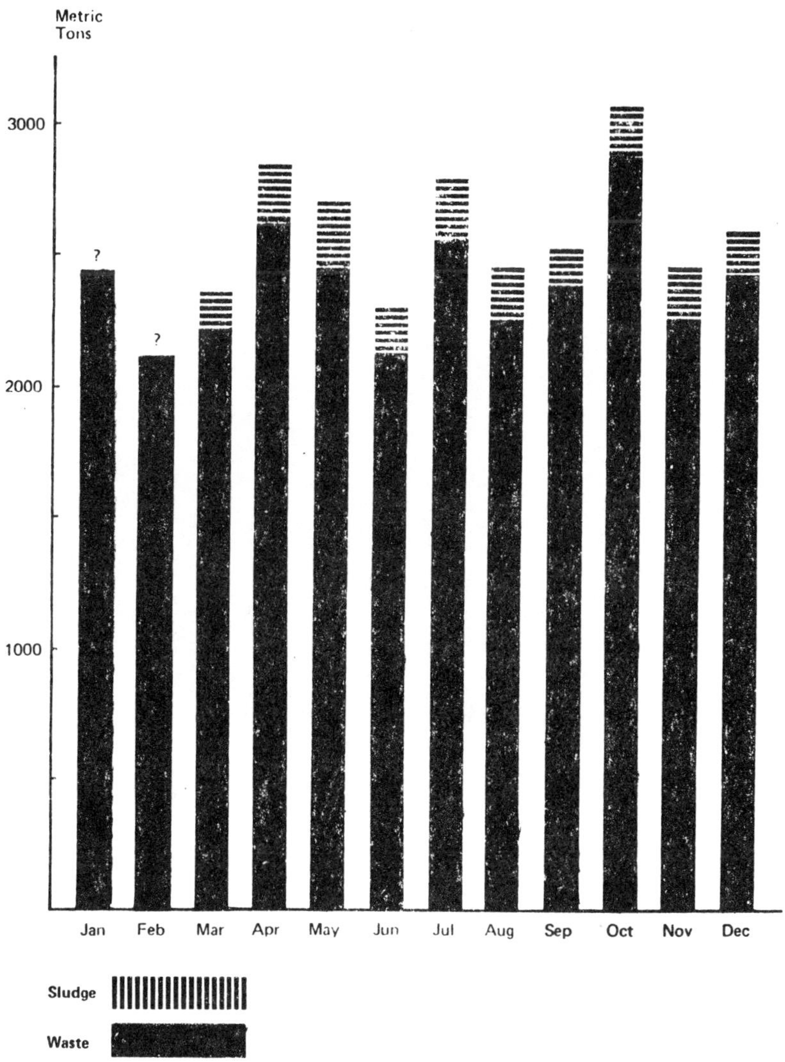

TABLE 6.5: QUANTITY OF WASTE INCINERATED (January-October 1976)

Operating Period	Quantity of Waste Incinerated (metric tons)*	Quantity of Sludge Incinerated** (metric tons)	Total Incinerated (metric tons)	Furnace Operations (hr)		Total (hr)	Average (metric tons/hr)
				1st Furnace	2nd Furnace		
January	2,567	195	**2,762**	464	460	**924**	3.0
February	2,289	272	**2,561**	426	426	**852**	3.0
March	2,694	215	**2,909**	488	488	**976**	3.0
April	2,586	261	**2,847**	463	463	**926**	3.1
May	2,435	228	**2,663**	451	450	**501**	3.0
June	2,808	238	**3,046**	517	519	**1,036**	2.9
July	2,920	177	**3,097**	533	513	**1,046**	3.0
August	2,910	182	**3,092**	516	517	**1,033**	3.0
September	2,917	208	**3,125**	530	530	**1,060**	2.9
October	3,100	154	**3,254**	538	538	**1,076**	3.0
Total	**27,226**	**2,130**	**29,356**	**4,926**	**4,904**	**9,430**	

SOURCE: INOR

* To convert to short tons, multiply by 1.1.
**At 40-50 percent moisture content.

In February 1975, the sewage treatment unit and the sludge dryers began operation. In that first year, 1,970 metric tons (2,170 short tons) of sludge were incinerated, while 2,100 metric tons (2,300 short tons) were burned during the first 10 months of 1976 (see Table 6.6). Accordingly, the total amount of incinerated solid waste and sludge was 1,300 metric tons (1,430 short tons) in 1973, 23,100 metric tons (25,400 short tons) in 1974, 31,070 metric tons (34,240 short tons) in 1975, and 29,350 metric tons (32,000 short tons) during the first 10 months of 1976. Theoretically, with two furnaces operating at full capacity (7,000 hr/yr) 49,000 metric tons (54,000 short tons) of waste could be treated. [The manufacturers and the municipality estimate that a minimum of 30,000 metric tons (33,000 short tons) per year is necessary to justify 24-hour-a-day operations. When operating around the clock, the plant is shut down at 4 AM on Saturday mornings. One week out of two it is started up again at 4 AM on Monday. In alternate weeks, Monday is used for routine maintenance and operations begin again Monday evening.] In 1976, approximately 37,000 metric tons (40,750 short tons) were incinerated—34,000 metric tons (37,400 short tons) of solid waste and 3,000 metric tons (3,300 short tons) of dried sludge. This amounts to 75% of maximum capacity, or almost 80% of the manufacturer's rated capacity of 43,000 metric tons (47,400 short tons) per year.

The increase in the quantity of waste treated is the result of new communities joining Brive's collection and treatment scheme. Indeed, at the rate of planned growth, the capacity of the incineration unit may be insufficient by the end of 1978 (14), and a third furnace will be necessary.

The energy content of the waste incinerated during the first 10 months of 1976 was $42{,}480 \times 10^6$ kcal ($165{,}580 \times 10^6$ Btu). The heat (energy) actually recovered and sent to the sludge dryers during that period was equal to $12{,}000 \times 10^6$ kcal ($47{,}600 \times 10^6$ Btu). This equals a 28% energy recovery rate, assuming the waste has a lower heating value of 1,500 kcal/kg (2,700 Btu/lb) (estimated by the plant operator; no measurement has been made). This figure varied from 18% to 49% per month, depending on the sludge drying requirement. Experience has shown that rural areas like Brive produce a significant amount of damp grass and leaves for processing which may create combustion problems and necessitate extending drying time on the first grate. However, the lower heating value of the waste has increased with the adoption of plastic bag collection in Brive, since the plastic bags not only protect the refuse from any further inclement weather but also produce intense heat when burned with the waste. (The heating value of waste at any incineration site will vary somewhat over time. Such variations may occur because the composition of the waste can vary considerably by season or as a result of economic conditions. For example, when the cost of paper increases, paper is recycled rather than discarded as waste. Since paper has a relatively high heat content its absence from the waste stream might result in waste of lower heat content.)

The saturated steam produced in the heat-recovery process is at 10.5 bars (10.4 atm) not the 16 bars (15.8 atm) stipulated in plant specifications. The energy content of the steam is assumed to be 560 kcal/kg (1,008 Btu/lb). The

TABLE 6.6: ENERGY PRODUCTION AND USE (January-October 1976)

Period of Operation	Quantity of Waste Incinerated (metric tons)*	Quantity of Sludge Incinerated (metric tons)	Energy Content of Waste (10^6 kcal)	Production of Steam** (metric tons)	Steam Use (10^6 kcal)†			Heat Recovery Rate‡ (percent)
					Sludge Drying	Air Preheating	Condensor	
January	2,567	195	4,001	7,228	955	1,345	1,748	24
February	2,289	272	3,643	6,000	1,790	1,240	329	49
March	2,694	215	4,207	6,290	1,514	1,421	587	36
April	2,586	261	4,080	5,556	1,526	1,348	237	37
May	2,435	228	3,828	5,413	1,114	1,312	605	29
June	2,808	238	4,395	6,092	1,167	1,509	736	27
July	2,920	177	4,516	5,884	970	1,113	1,212	21
August	2,910	182	4,505	5,874	1,005	1,099	1,185	22
September	2,917	208	4,536	5,874	1,142	1,128	1,019	25
October	3,100	154	4,769	6,182	846	1,145	1,472	18
Total	**27,226**	**2,130**	**42,480**	**60,393**	**12,029**	**12,660**	**9,130**	

SOURCE: RPA estimates based on INOR records.

* To convert to short tons, multiply by 1.1.

**Although the production of steam is measured constantly, these figures are based on the number of hours of actual operation.

† To convert kcal to Btu, multiply by 3.9685.

‡ The heat recovery rate equals the amount of heat actually used to dry sludge divided by the heat content of the waste.

amount of this used in the sludge dryers has varied considerably from month to month—a 2:1 ratio between February 1976 and October 1976—because of differences in the quantity and composition of the sludge. Thus, the amount of steam required to dry one cubic meter (1.3 cu yd) of sludge ranges from 1.6 metric tons to 2.3 metric tons (1.75 to 2.5 short tons), or an average of 1.8 metric tons per m^3 (1.5 short tons per cu yd) (15). The resulting dried sludge (with a moisture content of 40 to 50%) has had a lower heating value of 770 kcal/kg (1,390 Btu/lb), while the lower heating value of the solid portion of the sludge is estimated at 2,730 kcal/kg (4,900 Btu/lb) (16).

The average thermal efficiency of the incinerator and boiler is between 60% and 66%, depending on the lower heating value of the waste incinerated. (If the value is low, more energy is required for preheating combustion air.) A lower heating value of 1,300 kcal/kg (2,340 Btu/lb) gives a thermal efficiency of 60%, and a lower heating value of 1,600 kcal/kg (2,880 Btu/lb) or above gives a thermal efficiency of 66% (17). This level of efficiency could be improved if a better balance could be achieved between the amount of waste incinerated and the amount of sludge treated. As it is, the amount of energy produced from incineration exceeds the amount of energy needed for sludge drying. Brive has plans to build municipal greenhouses near the waste treatment complex and heat them with the steam not used for drying sludge.

However, a thermal balance over a period of a year does not ensure a balance on every single day. At Brive, sludge must remain in the thickener near the limit of capacity over the weekends when no waste is incinerated.

In terms of reliability, the system performed well during the early years of operation. (See Figure 6.6 for the manufacturer's guarantee on the parts and operation of the incinerator.) Specifically, during the first 3 years of operation, nonroutine maintenance costs amounted to FFr 441,000 ($92,260; 1976), including FFr 131,000 ($27,400; 1976) to replace the multicyclones after 12,000 hours of operation, or FFr 5.2 per metric ton ($1 per short ton) incinerated (see Table 6.7). The manufacturer claims nonroutine maintenance costs generally run FFr 8 to 10 per metric ton ($1.50 to $1.90 per short ton; 1976) (18). And during roughly the first 2 years of operation, waste incineration had to be suspended for only 634 hours (see Table 6.8) (19).

However, by 1976 the refractory walls of each furnace had been severely damaged. Approximately 18 m^2 (194 sq ft) of refractory material will have to be replaced in the upper part of each furnace. According to the municipality, the expansion joints failed to prevent distortion of the furnace walls at high temperature (2). According to the manufacturer, the problems have occurred because the sludge filtration step is partly bypassed and sludge enters the dryers with a higher moisture content than anticipated. In addition, evaporated water is circulated to the furnaces to eliminate odors, and thus the furnaces receive more water vapor when the sludge dryers are operating than had been expected.

FIGURE 6.6: INCINERATOR MANUFACTURER'S GUARANTEE

Von Roll guarantees its furnaces under the following provisions:

- The lower heating value of the waste incinerated must be between 1,000 and 2,500 kcal/kg (1,800 to 4,500 Btu/lb).
- The moisture content of the waste must be under 50%.
- The carbon content of the waste must be under 5%.
- The ash content of the waste must be under 40%.
- The combustible content of the waste must be over 25%.

Under these conditions, Von Roll guarantees that:

- Each furnace will have a capacity of 3.5 metric tons (3.9 short tons) per hour.
- Putrescible materials in the residues of incineration will be under 0.5% by weight.
- Unburned materials in the residues of incineration will be under 5% by weight.
- The temperature of incineration will be at least 700°C (1292°F).
- The particulate content of the stack gas will not be more than 600 mg/m^3 at 7% CO_2.
- The CO content of the gas will not be more than 0.5%.
- The electricity required for incineration will not be more than 40 kWh per metric ton (36.3 kWh per short ton) of waste.

Mechanical equipment is guaranteed for one year.

TABLE 6.7: NONROUTINE MAINTENANCE COSTS (as of December 1976)

	FFr	1976 $
Replacement parts for grates, ash chain conveyor, and furnaces	217,500	45,500
Replacement of multicyclone		
–for Furnace 1	75,000	15,690
–for Furnace 2	56,000	11,715
Replacement of refractory	92,500	19,355
Total	**441,000**	**92,260**

SOURCE: INOR

TABLE 6.8: INCINERATOR BREAKDOWNS

Cause	Downtime (hours) 1974	1975
Steam network failure	22	8
Crane failure	70	77
Grate failure	21	37
Boiler failure	30	78
Multicyclone failure	30	57
Hopper failure	22	17
Electrical circuit failure	16	5
Ash extractor failure	10	6
Lack of water	12	6
Lack of electricity	11	14
Masonry failure	26	47
Lack of waste	—	12
Total	**270**	**364**

SOURCE: Municipality of Brive.

But regardless of the cause, it will cost about FFr 500,000 ($105,000; 1976) to replace the damaged sections of the refractory walls. With this repair, the cost of nonroutine maintenance will amount to 10% of the cost of the incinerator.

Finally, after 3 years operating experience, it appears the environment of the complex is quite acceptable. Although there is some odor in the vicinity of the sludge dryers, it does not spread to the rest of the plant. Moreover, with the help of outside specialists, the operational noise that was inaudible at the plant but could be heard several miles away has been eliminated. (A fan produced a peculiar resonance.) And, very importantly, no major accident has ever occurred at the Brive facility.

COSTS OF THE SYSTEM

The total cost of the complex, exclusive of the land which belonged to the municipality, was FFr 30 million ($6 million; 1971–1974): FFr 11.2 million ($2.2 million; 1971–1974) for the incineration unit, FFr 14 million($2.85 million; 1971–1974) for the sewage treatment unit, and FFr 4.8 million ($950,000; 1971–1974) for the sludge dryers. Government subsidies paid for 50% of the cost of the incineration unit: the Ministry of the Interior provided 30%, and the Department of Corrèze provided 20%. For a while, the law permitted the recovery of some of the taxes because the complex was managed by a private company, but this is no longer the case. Subsidies for the sewage treatment unit and sludge dryers amounted to 55% of the costs—a 30% subsidy from the Ministry of Interior and a 25% subsidy from the local water pollution agency of the Adour-Garonne Basin (Agence Financière de Bassin Adour-Garonne).

The rest of the funding was obtained through 20-year loans, primarily from the Caisse des Dépôts et Consignations, which plays the role of state banker for such projects at relatively low interest rates (7.8% in this case). In addition Société Générale provided a FFr 1.7 million bank loan at 8.05%. Thirty-seven percent of the cost of the incinerator unit was expended in 1972; 55%, in 1973; 7%, in 1974; and 1%, in 1975.

The overall cost of incineration in 1975 was FFr 59.5/metric ton ($12.60 per short ton; 1975) incinerated, based on approximately 30,000 metric tons (33,000 short tons) incinerated annually. The cost is composed of the capital cost of FFr 27.5/metric ton ($5.83/short ton; 1975) (based on amortization cost, including debt service, of FFr 800,000 over 20 years) plus the direct operating cost of FFr 32/metric ton ($6.77/short ton; 1975).

Capital Costs

Although the original contract for the incineration unit stipulated a capital cost of FFr 9.5 million ($1.7 million; 1971) (see Table 6.9), Brive eventually paid FFr 11.2 million (including a value added tax at the rate of 17.6%) for the unit because of labor and material cost increases. This "overrun" was allowable under an escalator clause in the contract and the contract was revised according to an index for construction costs.

TABLE 6.9: COST OF INCINERATION UNIT (1971 prices)

	FFr	FFr	1971 $
Construction		2,340,000	424,700
Equipment		5,965,100	1,082,600
–Scale	58,800		
–Waste pit doors	84,700		
–Overhead crane	318,000		
–Ash extractor	282,200		
–Furnaces	3,082,300		
–Steam boilers	764,400		
–Cyclone battery	515,100		
–Control equipment	224,600		
–Electrical equipment	382,200		
–Other equipment	252,800		
Access roads		369,500*	67,000
Administration building		155,000*	28,300
Add-on for increase in capacity of furnaces		658,000	119,400
Total		**9,487,600****	**1,722,000**

SOURCE: INOR; City of Brive.

* Fifty percent of total cost; remaining 50 percent allocated to waste-water treatment unit costs.
**Including a 17.6 percent value added tax.

The sewage treatment unit cost FFr 14 million ($2.5 million; 1971) (including a value added tax at the rate of 17.6%). Of this, FFr 800,000 ($145,000; 1971) was spent on the sludge dryers, excluding installation.

Direct Operating Costs

Brive pays INOR four kinds of waste-incineration fees, the amounts of which are based on the quantity of waste to be incinerated and, consequently, on the number of shifts required, as shown in the following matrix.

INCINERATION FEES PAID TO INOR
(rates in effect in March 1975)

	Quantity of Waste/Number of Shifts	
Type of Fee	15,000 to 30,000 metric tons/2 shifts per day	30,000 to 43,000 metric tons/3 shifts per day
A fixed fee for operation and routine maintenance	FFr 424,000/annum ($98,800/annum)	FFr 832,000/annum ($193,900/annum)
An additional fee for operation and routine maintenance proportional to the tonnage of waste treated	FFr 10.47/metric ton ($2.22/short ton)	FFr 14.93/metric ton ($3.16/short ton)
A fixed fee for nonroutine maintenance and replacement parts	FFr 39,000/annum ($9,000/annum)	FFr 136,000/annum ($31,700/annum)
An additional fee for nonroutine maintenance and replacement parts proportional to the tonnage of waste treated	FFr 3.24/metric ton ($0.69/short ton)	FFr 3.24/metric ton ($0.69/short ton)

This fee schedule was established in the original contract and is revised quarterly, with an average increase of approximately 1% per month.

In 1975, approximately 31,000 metric tons (34,000 short tons) of waste were incinerated: 27,400 metric tons (30,200 short tons) of household waste, 1,700 metric tons (1,870 short tons) of commercial and industrial waste, and 1,900 metric tons (2,090 short tons) of dried sludge. The cost of the incineration of household waste and dried sludge, as billed by INOR, was FFr 1.5 million ($350,000; 1975), or FFr 51.2/metric ton ($10.83/short ton; 1975) incinerated: fixed fees of FFr 968,000 and proportional fees of FFr 532,000. The actual cost to the city was somewhat less, as INOR collects FFr 48/metric ton ($10.15 per short ton; 1975) of industrial waste received from private generators and returns FFr 6/metric ton ($1.25/short ton; 1975) to the city. Thus, in 1975, the city received an additional FFr 10,200 ($2,375; 1975) which should be deducted from the fees it pays INOR. The actual cost to the city is therefore FFr 50.8/metric ton ($10.75/short ton; 1975). Brive paid the fixed and proportional fees directly to INOR and then allocated them among the municipalities participating in the collection/incineration program, according to the quantity of waste collected in each community.

If this heat were not available, the municipality would have to purchase an equivalent amount of energy in order to dry sludge. Therefore, the value of the heat recovered from the incineration process and used in the sludge dryers should be credited towards these costs. For example in 1975, an estimated 22,500 metric tons (24,800 short tons) of steam were used for drying sludge. Based on a price of FFr 25/metric ton ($5.30/short ton; 1975) of steam, this amounts to FFr 560,000 ($130,500; 1975) or FFr 19.1/metric ton ($4.04 per short ton; 1975) of waste incinerated (21). This would have reduced fees to the municipalities to FFr 32/metric ton ($6.77/short ton; 1975) of sludge and household waste.

Brive also pays INOR an annual fixed fee of FFr 133,000 ($27,800; 1976) for operating the sludge dryers and a proportional fee of FFr 0.0056/m^3 ($0.0009 per cu yd; 1976) of sludge treated. This cost excludes the price of the steam which is obtained at no additional cost.

REFERENCES

Notes

(1) Private communication: Mr. Val, director of the technical services of the Municipality of Brive.
(2) Private communication: Mr. Allemandour, engineer, technical services of the Municipality of Brive.
(3) Municipal records: *Devis programme de concours pour la construction d'une usine d'incinération et d'une station d'épuration.*
(4) Municipal records: *Procés verbaux des réunions successives du jury de concours.*
(5) RPA estimates based on INOR operating records.
(6) Municipal records: *Devis descriptif équipement-incinération.*
(7) ODA: *Ville de Brive, Station d'épuration.*
(8) M. Krings (INOR), *French Experience with Facilities for Combined Processing of Municipal Refuse and Sludge.*
(9) Private communication: Mr. Gargallo, chief of operations of the complex (INOR).
(10) Private communication: Mr. Allemandour.
(11) INOR operating records.
(12) Municipal records: *Mémoire technique—incinération.*
(13) Private communication: Mr. Arbruster, director, INOR.
(14) Private communication: Mr. Val.
(15) INOR operating records.

(16) Private communication: Mr. Arbruster.
(17) Private communication: Mr. Arbruster.
(18) Private communication: Mr. Arbruster.
(19) INOR operating records.
(20) Private communication: Mr. Val.
(21) RPA estimates based on INOR operating records.

Persons Contacted

Mr. Val	Director of Technical Services, Municipality of Brive
Mr. Allemandour	Principal Engineer, Technical Services, Municipality of Brive
Mr. Arbruster	Director, INOR
Mr. Gargallo	Chief of Operation of the waste treatment complex, INOR
Mr. Amouroux	Operator of the sewage treatment unit, ODA

An Overview

Material for this chapter has been based upon a report prepared by Resource Planning Associates, Inc., Washington, D.C. and Paris (CONS/2103-6).

EUROPEAN WASTE-TO-ENERGY SYSTEMS

The recovery of energy through the combustion of municipal solid waste is a well established technique for conserving energy. Combustion units can produce electricity; hot water for domestic use; and steam for district heating, industrial processes, or the drying of sewage sludge. A combustion unit is a facility built at one time in a single location. One building (or plant) may thus contain more than one unit and one unit may contain more than one furnace. A single city may have several combustion plants, each housing several units. These units may be radically different in design, and therefore are considered separately.

There are fewer than 20 such waste-to-energy systems in the United States. In western Europe, however, 243 combustion units are currently recovering energy from municipal solid waste. The oldest of these facilities went into service before World War II. Tables 7.1 and 7.2 present some key facts about Western European countries and their waste-to-energy units.

If the use of this technique is to spread in this country (and the needs to conserve energy and protect the environment suggest that it should) the lessons of Western European systems can be valuable to those planning for similar systems in the United States. The Buildings and Community Systems Branch of ERDA'S Office of Conservation therefore engaged Resource Planning Associates, Inc. (RPA) to make case studies of representative waste-to-energy

TABLE 7.1: AREA, POPULATION, AND POPULATION DENSITY OF WESTERN EUROPEAN COUNTRIES

	Area (thousand km²)	Population (thousands of inhabitants)	Density (inhabitants per km²)
Austria	84	7,528	90
Belgium	30	9,804	327
Denmark	43	5,045	117
Finland	337	4,682	14
France	547	52,507	96
Greece	132	8,962	68
Ireland	70	3,086	44
Italy	301	55,361	184
Luxembourg	3	342	132
Netherlands	41	13,541	330
Norway	324	3,987	12
Portugal	92	8,735	95
Spain	505	35,225	70
Sweden	450	8,161	18
Switzerland	41	6,481	158
United Kingdom	244	55,933	229
West Germany	249	62,041	249

SOURCE: INSEE (French National Institute of Economic Statistics).

TABLE 7.2: WASTE-TO-ENERGY SYSTEMS IN WESTERN EUROPE

Country	Number of Units*	Number of Plants**
Austria	2	2
Belgium	3	3
Denmark	45	31
Finland	2	2
France	29	20
Italy	16	14
Luxembourg	1	1
Netherlands	7	6
Norway	13	8
Spain	3	2
Sweden	22	16
Switzerland	33	29
United Kingdom	9	9
West Germany	58	38
Total	**243**	**181**

* A unit is a facility built at one time in a single location.

**A plant is the building in which one or more waste-to-energy units is installed.

systems in Western Europe. These case studies examine the technical and economic aspects of combustion units in detail.

This overview discusses some common features of the Western European systems. First, the combustion techniques available and the end uses to which the recovered energy is put are considered. Next, the economics of the systems are discussed briefly. Finally, a census of the systems currently operating or under construction in Western Europe is presented.

SYSTEM CONFIGURATIONS

Waste-to-energy systems generally consist of a furnace, heat recovery equipment (usually a boiler), and some system for making use of the energy recovered (e.g., a steam pipe network or a turbo-generator). Most systems have some form of air pollution control. A few have shredders to process the municipal solid waste before it enters the furnaces. In France, for example, only 4 of the 29 combustion units use shredders.

Furnaces

The furnace grate which agitates and transports the waste during combustion is the essential part of a combustion unit. These grates must be able to accept the wide variety of objects and substances which make up municipal solid waste.

Five manufacturers account for 85% of the grates installed in Western Europe.

The breakdown is as follows:

Martin (West Germany)	14%
Vereinigte Kessel Werke (West Germany)	16%
Von Roll (Switzerland)	24%
Bruun & Sorensen (Denmark)	18%
Volund (Denmark)	13%
Other	15%

The market share of each manufacturer has remained relatively constant, except for Volund which has installed fewer systems in the last five years. The share of Bruun & Sorensen would be smaller if capacity were being measured since their system is used mostly for small units.

Among the other West European manufacturers are:

- Bartholomeus (Italy)
- Carbonisation Entreprise et Céramique (France)
- Lurgi (West Germany)
- Nichols (U.K.)
- Alberti Fonsard (Italy)

- Sulzer (Switzerland)
- Kohlen Scherdungs Gessellschaft (West Germany)
- Heenan (U.K.)
- Stein (France)
- Lokomo (Finland)
- Esslingen (West Germany)
- Destructor (Sweden)
- Kraus-Maffei (West Germany)
- Venien (France)

No single grate system seems best for all types of combustion units. For example, the VKW Design is used mainly in large units, while the Bruun & Sorensen design is used mainly in small units. There have been small modifications but no basic changes in grate design. The tendency seems to be for each firm to improve the flexibility and responsiveness of grates so that they can adapt to changes in waste characteristics. Some of the principal grate designs are described in the following pages.

Alberti System: The single-section grate is composed of fixed and moving elements arranged in steps. Combustion takes place on the upper bars which are fixed, while the lower moving bars insure the agitation of the waste:

FIGURE 7.1: ALBERTI SYSTEM

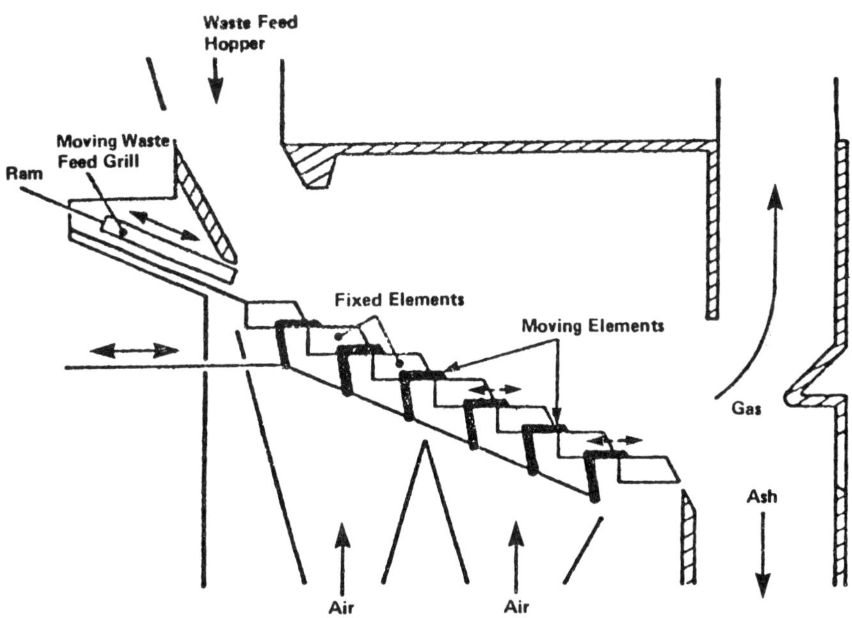

Bruun and Sorensen System (Denmark): The grate has two or three sections with six horizontal axles in each. Grate prongs are screwed to the axles. Transport and agitation are assured by means of a hydraulic system causing axles no. 1, 3 and 5 in each section to turn clockwise while no. 2, 4 and 6 turn counterclockwise.

FIGURE 7.2: BRUUN AND SORENSEN SYSTEM

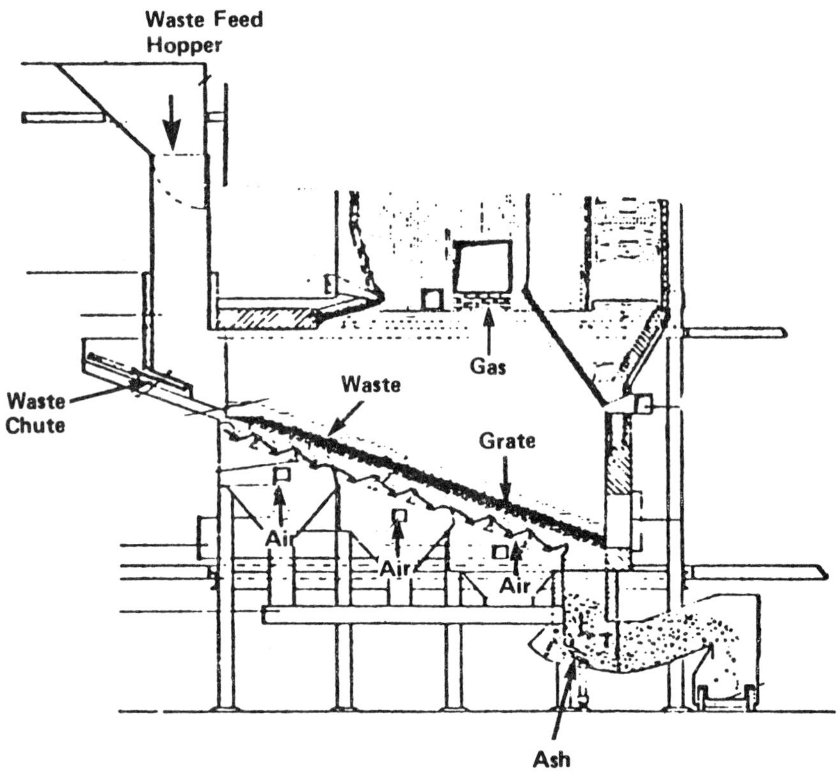

Carbonisation Entreprise et Céramique (CEC) System (France): The grate has one section, but there may be two or more grates arranged in parallel. The grate is composed of successive sliding, rocking, and fixed elements. The sliding and rocking elements are synchronized so that the sliding elements move over the rocking elements when these are retracted. By this movement, the sliding elements insure the transport of the waste and the rocking elements insure the agitation.

FIGURE 7.3: CEC SYSTEM

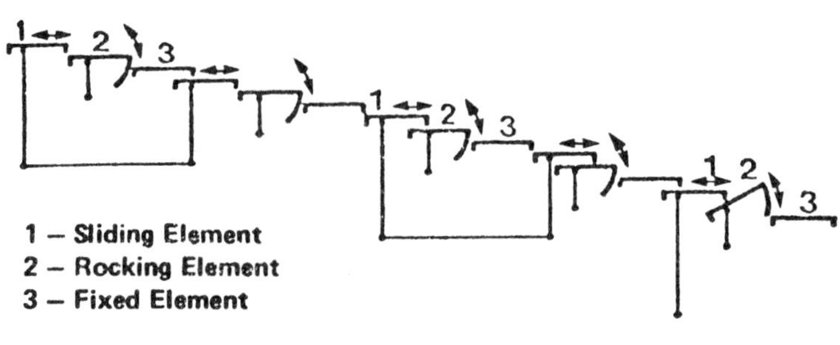

Esslingen System (West Germany): The single-section grate is composed of semicircular rocking elements. The movement of these elements insures transport and agitation of the waste. Each element is made of bars placed side by side through which combustion air circulates.

FIGURE 7.4: ESSLINGEN SYSTEM

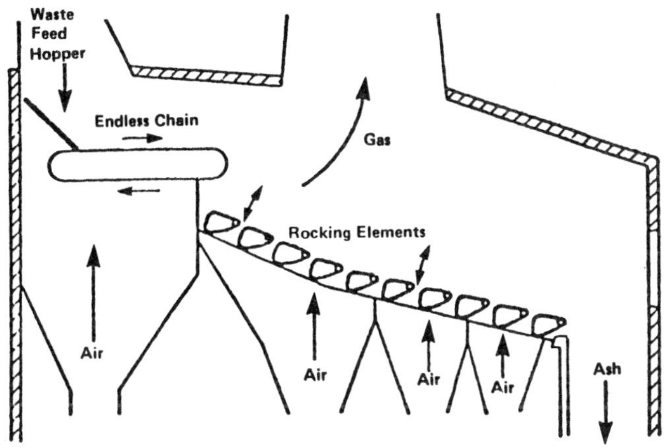

Heenan Nichol System (U.K.): The grate is composed of three or more sections arranged in steps. The first section distributes and dries the waste, the following sections are used for combustion. Each section is composed of rocking elements which insure the agitation of the waste. Each pair of elements moves reciprocally so that at any moment, half of the elements are moving. The odd numbered elements are linked and move in unison, followed by the even numbered elements, which are also linked.

FIGURE 7.5: HEENAN NICHOL SYSTEM

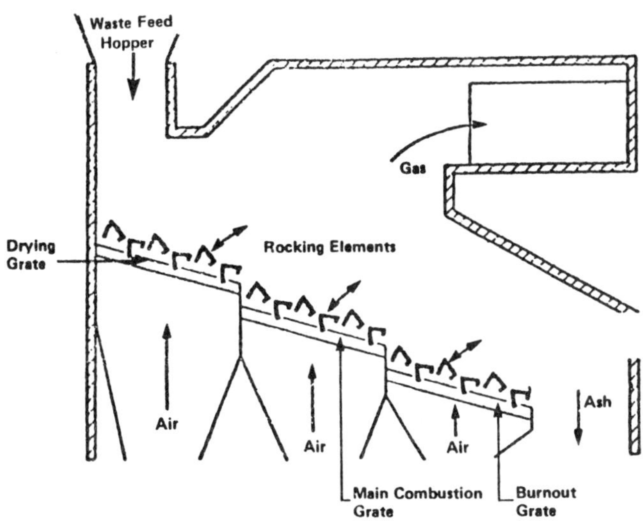

Martin System (West Germany): The grate is composed of a single section, but there may be more than one grate in parallel alignment. Each grate is composed of alternate fixed and moving bars, the moving bars sliding up over the fixed bars. The reverse action of the grate insures a continuous agitation of the waste as it falls under gravitational force. The bars contain passages so that air circulates within the bars and enters the combustion zone through slits between the bars.

FIGURE 7.6: MARTIN SYSTEM

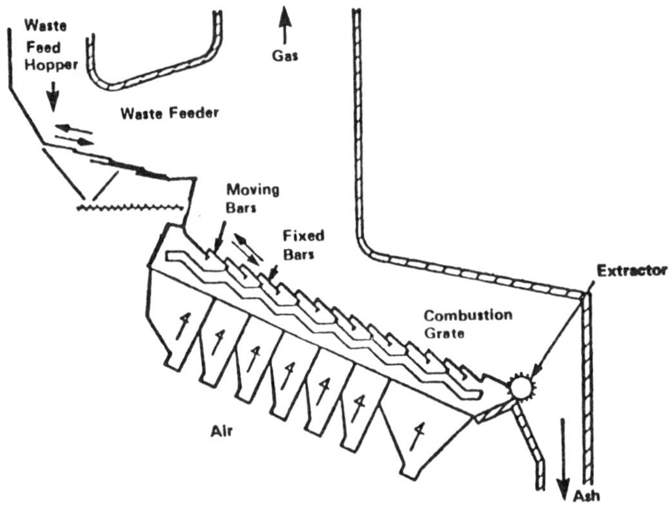

Vereinigte Kessel Werke (VKW) System (West Germany): The grate consists of rigid rollers, arranged in series with horizontal axes. The roller bodies are faced with cast iron grate bars. The speed of the individual rollers is variable. Combustion air enters through the grate bars of the rollers and can be regulated for each roller.

FIGURE 7.7: VKW SYSTEM

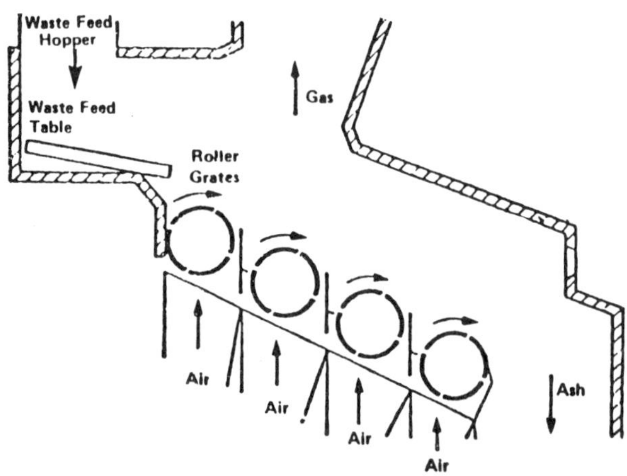

Volund System (Denmark): This system is similar to the Von Roll system. The grate is in two sections, the first for drying the waste, the second for combustion. Instead of a third section, there is a rotary kiln to insure a complete burnout of the waste.

FIGURE 7.8: VOLUND SYSTEM

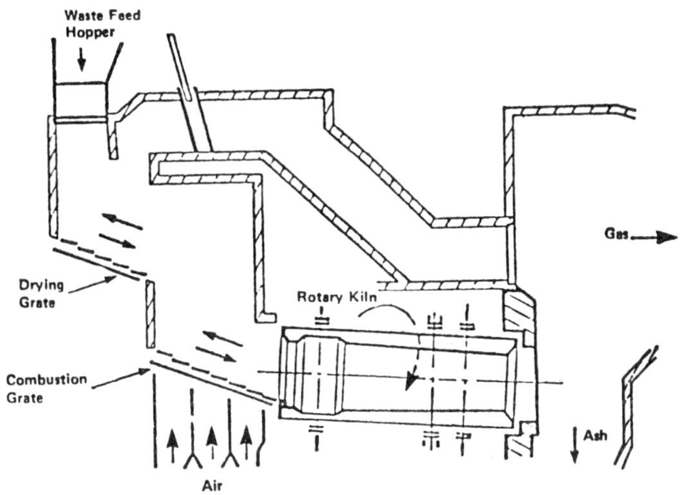

Von Roll System (Switzerland): The grate is composed of three sections, the first for drying and igniting the waste, the second for combustion, and the third for burnout. Each section is composed of parallel bars, alternately fixed and moving, in a single row. The moving bars have a horizontal action, insuring the agitation of the waste. Combustion air is introduced through holes in the front and sides of the bars.

FIGURE 7.9: VON ROLL SYSTEM

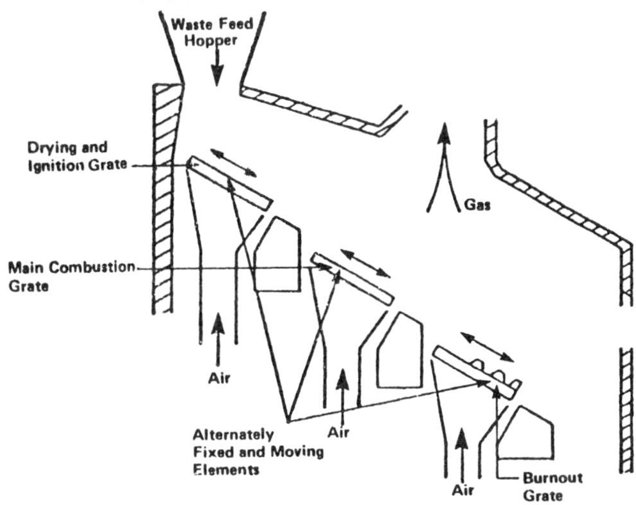

Heat Recovery Equipment

In nearly all the combustion units, energy is recovered through the medium of steam. In practice, this can mean either the classic fire-tube boilers or the newer waterwall boilers. Since the furnace and boiler form a closed system, it is possible to calculate a thermal energy balance and determine the percentage of the heat content of the waste introduced into the furnace that is actually recovered as usable energy. Sample thermal energy balance sheets are shown in Tables 2.2 (page 49) and 5.7 (page 146).

A few combustion units (e.g., Tergnier in France) do without a boiler and use the hot combustion gases directly for the drying of sewage sludge.

End Use

The main uses of the recovered energy are to generate electricity; to heat water for domestic use; and to produce steam for district heating, an industrial process, or the drying of sewage sludge. A number of other uses (e.g., heating swimming pools and greenhouses) have been tried or suggested. The following diagram shows one typical way of using recovered energy.

FIGURE 7.10: PLAN FOR USING RECOVERED ENERGY

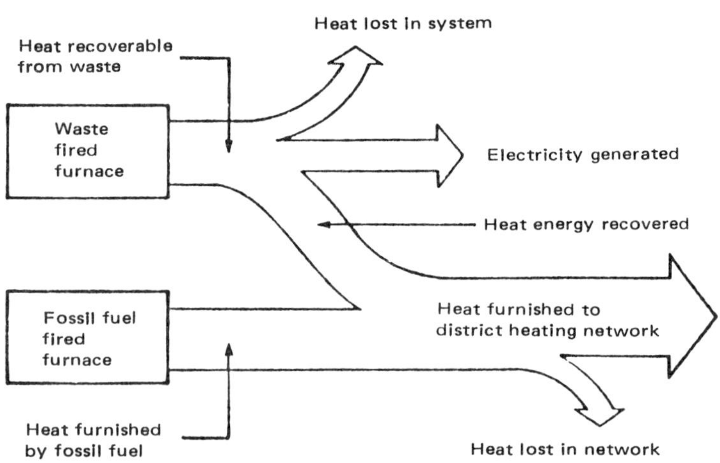

Electricity can be produced with a condensing turbo-generator (e.g., the Cheneviers unit at Geneva, Switzerland) which uses all of the steam available. It can also be cogenerated, along with low pressure steam, using a back pressure turbo-generator (e.g., as in Munich, West Germany). However, in Europe, the production of electricity is sometimes a state monopoly and, in any case, is subject to many regulations. Institutional barriers are therefore a serious concern in the production of electricity from solid waste. For example, the high cost of generating electricity from waste (as the case studies of Munich, Germany and Geneva, Switzerland indicate) may be due as much to institutional as technical reasons.

The use of steam for domestic water heating or district heating generally works best in a new town built in conjunction with the waste-to-energy system. The steam pipe (or hot water pipe) network can then be laid down during initial construction and the community planners can incorporate the heating system into their design.

When sewage sludge is dried, either directly by the combustion gases or indirectly in steam jacketed driers (see the case study of Brive, France), the dried sludge can be incinerated along with the solid waste. Such a system requires that the combustion unit be built in conjunction with a wastewater treatment facility.

Whatever the end use of the recovered energy, there may well be a conflict between its production and the other purpose of the plant—reduction of the volume of municipal solid waste. If the operators see this latter as their overriding purpose, they may choose operating trade offs which maximize availability rather than the efficient production of energy.

Matching of the unit with the load demand of the energy user may often be difficult. Peak demands for electricity, for example, may not coincide with the arrival of the greatest quantities of waste (see case study of Munich). The capacity of the waste bunker to hold a supply of several days' waste may be crucial.

Pollution Control

Since most Western European waste-to-energy systems are located in urban areas, pollution control is important and becoming more so. Regulations vary from country to country, which means that different pollution equipment is required.

West Germany has the strictest laws in Western Europe. Stack gases may have no more than 100 mg of particulates per cubic meter. (In the United States, the national standard for incinerator emissions permits a maximum of 180 mg/m^3 of particulates. Some states and cities have stricter limits.) There are limits on emissions of carbon monoxide, chlorine compounds, and fluorine compounds. Since 1974, when these regulations went into effect, new combustion units in West Germany have required electrostatic precipitators followed by at least partial scrubbing.

In other countries, only particulate emissions are regulated, and scrubbers are not needed. In France, for example, the limits for particulate emissions are 600 mg/m^3 for furnaces with a capacity between 1 and 4 metric tons/hour; 250 mg/m^3 for furnaces with a capacity between 4 and 7 metric tons/hour and

150 mg/m^3 for furnaces with a capacity about 7 metric tons/hour. Finally, in some countries, such as Denmark, the regulations are lax enough that most incinerators currently need only multicyclones. It is clear, however, that the trend in all Western European countries is toward stricter regulations of particulate emissions. This means that electrostatic precipitators will be required in nearly all cases. These devices will mean higher capital costs, but no basic technological changes in the waste-to-energy systems.

The use of scrubbers is not widespread. Although the equipment exists, it is very expensive and needs frequent maintenance. It is not certain that, at the moment, any plant operates with continuous scrubbing. For this reason, most countries are hesitant to impose strict legislation until improved scrubbers appear.

Technical Problems

The techniques commonly used to recover energy from municipal solid waste in Western Europe are mature and stable. Most combustion units have good operating records. Most improvements have been in response to changes in the characteristics of the waste (e.g., more plastics) and stricter pollution control requirements.

Boiler corrosion has been the single most pervasive problem. To understand the effect of this problem on design, consider the experience of the Von Roll company of Switzerland over the past 25 years.

The first basic Von Roll combustion units were designed with a refractory firebox and a second chamber housing the boiler. This was a good, reliable design, and had a very good operating record. The first unit of this generation came on stream in 1954 at Berne, Switzerland. Several more such plants were built in Switzerland and elsewhere in Europe, and Von Roll's reputation was established as a successful waste combustion design company.

However, these plants were large and costly. Since they generally produced only saturated steam, much of the heat was wasted. By the late 1950s and early 1960s, there were complaints that the design was not good enough.

Thus, a second generation was designed. In the new unit, the boiler was moved forward right into the firebox. These integrated units incorporated waterwall tubes brought down near the combustion grates and superheater tubes to generate superheated steam. The new units were smaller and cheaper, and promised to heat much more efficiently.

The first unit of this generation was ordered in 1962, for Winterthur, Switzerland. Operation was to begin in 1965. The second unit, also ordered in 1962, was for Frankfurt, West Germany. The third was for Geneva, Switzerland. These two units were to begin operating in 1966. Altogether, a total of twenty-seven such units were ordered by the end of 1968 when the first real evidence of major problems appeared.

Corrosion of various kinds began to appear rather early in the operating life of the first of these new units. By the end of 1968, it was obvious that some design changes were needed. The first and most important corrosion noted was on the lower portion of the frontal and lateral waterwall tubes. Von Roll eventually decided that this was due to several factors. First, combustion was not completed on the burning grate but continued in small particles and in volatile matter to a considerable height. Thus, flames were much higher than anticipated. Second, since this combustion situation was not continuous, there was a constant shifting from an oxidizing to a reducing atmosphere in the furnace. This caused severe corrosion on the metal tube walls.

Several remedies were used. The lower portions of the waterwall tubes were coated with a ceramic and refractory protective material. A secondary air source was also added above the grate to generate turbulence in the combustion air and thus to provide for complete burnout lower in the firebox. The net effects were decreased thermal efficiency and increased air flow through the system. These, in turn, required the installation of a larger induced draft fan. The electrostatic precipitators also proved inadequate to handle the greater gas flow.

Some of the corrosion was due to high temperature at the surface of the superheater tubes. The only solution for this problem was to run the furnaces at a somewhat slower rate, and this led to an effective reduction of capacity.

Von Roll's previously excellent reputation suffered as a large number of combustion units began to experience the same problems. Essentially, too many furnaces with the same defects had been built too soon.

Von Roll's answer was threefold. First, for the integrated furnaces, all of the superheater and economizer bundles were moved to the third passage and the refractory protection and secondary air systems in the firebox were redesigned. The waterwalls were kept, but the major corrosion problems were reduced.

Von Roll began to update the older refractory furnaces with tail-end boilers. This redesign resulted in a new type of unit, the first of which was ordered by Fürth and the second, shortly thereafter, by Landshut in West Germany. These units have refractory fireboxes and tail-end boilers, but produce superheated steam. A new system for cleaning the tubes by physically knocking the tube bundles to shake off soot was incorporated. The extraordinary success of this design is shown by the fact that the Landshut I furnace has operated for 28,374 hours (as of August 1976) without ever having to be shut down for cleaning the tubes.

Third, another new generation of furnaces was designed in the early 1970s. These units have waterwalls, but the rest of the tube bundles are arranged so that they can also be cleaned by knocking. The first of these units began operations in Quebec, Canada in 1974. Units of this design are now operating in

Saugus, Mass. (near Boston) and Dijon, France. Dijon is the only plant of this design generating electricity; the others recover heat through steam.

According to Von Roll, the big advantage of the refractory furnaces is their reliability. The plant at Lausanne, Switzerland which is now nearly eighteen years old, is still available nearly 90% of the time. The best that the early integrated boiler designs have been able to achieve is about 75%, and the best that Von Roll will guarantee in a contract, even for their newest integrated boiler design, is 80%.

As the example of Von Roll shows, there are two opposing tendencies in the design of combustion units. One is to put the boiler right in the firebox, with waterwall tubes brought down near the burning grate. The other is to locate the boiler away from the refractory firebox in a second chamber. The first design increases the efficiency of energy conversion and reduces capital costs. It seemed in recent years to be the better option. However, because of corrosion problems, the second option, which was never rejected by such manufacturers as Volund or Bruun & Sorensen, may have a second life.

ECONOMICS

The economics of a waste-to-energy system can be considered under two headings. The first is the choice of combustion rather than some other treatment for solid wastes. The second is the addition of energy recovery equipment if combustion has been chosen. The considerations are quite different.

The chief alternative to combustion for municipal solid waste is landfilling. Western Europe has historically had high population densities and, as a result, a shortage of land for disposal of solid waste. Thus, the costs of the chief alternative to combustion have been higher there than in the United States.

As the United States population density increases, this difference will narrow. If landfill sites can only be found at greater distances from the generators of waste, growing transportation costs will further narrow this difference. The trend to require accounting for environmental costs may well close the gap entirely. As a result, more and more American communities will choose combustion for the disposal of their solid waste.

In this case, the question will arise, is it worthwhile to recover the energy produced in the combustion. The answer depends upon the availability of an appropriate end use for that energy.

Direct use of heat energy avoids the inevitable losses in the generation of electricity. Burning waste may be a very expensive way to produce electricity.

If the heat can be used for domestic hot water, district heating, or an industrial process near to the combustion plant, the economics may be more favorable.

The recovery of energy from waste can be most cost-effective if the end use is well integrated with the combustion unit (and even the waste collection system). Increasing fuel costs can only reinforce this trend. For example, the solid waste of a Western European town of 150,000 inhabitants can provide district heating for 10,000 housing units. If those housing units and the combustion unit were part of a well integrated plan, the economics might be quite favorable (see the case study of Toulouse-Le Mirail).

The recovery of materials from the residues of combustion may slightly enhance the cost-effectiveness of waste-to-energy systems. Many plants recover ferrous metals for sale as scrap.

In France, the Bureau des Recherches Géologiques et Minières (B.R.G.M.), the equivalent of the U.S. Bureau of Mines, has built a pilot plant to recover glass, ferrous metals, aluminum, and other nonferrous metals from the residues of incineration. This plant has operated satisfactorily since 1974 and a large scale system is contemplated for Paris (Ivry). However, the economics are not very favorable under present market conditions and it is feared that the revenues will not be sufficient to amortize the investment. Table 7.3 summarizes the economics of several units examined in RPA's case studies.

CENSUS OF WESTERN EUROPEAN COMBUSTION UNITS

The number of waste-to-energy systems now in service in Western Europe is an indicator of the soundness of this technology. There has been tallied 243 units (located in 181 plants or buildings) in service, under construction, or planned for the near future.

Luxembourg, Denmark, Switzerland, the Netherlands, Sweden, and West Germany make the most use of combustion with energy recovery. These countries have more than 20 tons of hourly capacity per million inhibitants, or enough to recover energy from the waste produced by 40% of the population (see Table 7.4).

Several factors seem to have encouraged the use of this technology in these countries. They all have concentrated population, their cities are reasonably wealthy, and they have strong manufacturing industries (in particular, the major firms licensing grate design are headquartered in these countries).

Furthermore, they benefit from favorable conditions for the sale of energy. They all have cold weather, which makes district heating attractive, and local electricity producers (unlike the state monopolies of France or Italy), which makes the sale of electricity easier. Costs of alternative means of disposal, such as landfills, are high. Belgium, Spain, Austria, and Finland have very few units, and Greece and Portugal, none.

The combustion units are usually located in large cities which often have

TABLE 7.3: COST SUMMARY OF SAMPLE WASTE-TO-ENERGY SYSTEMS

(1975 unless otherwise noted)

City and Population	Number and Size of Units	Tonnage of Waste Incinerated Annually	End Use of Recovered Energy	Capital Cost	Direct Operating Cost	Total Cost
Toulouse, France 400,000	3 x 8.15 metric tons/hr (3 x 9 short tons/hr)	100,000 metric tons (110,000 short tons)	District heating, domestic hot water, electricity for use in plant	FFr 34/metric ton* ($6.80/short ton)	FFr 7/metric ton* ($1.40 short ton)	FFr 41/metric ton* ($8.20/short ton)
Brive, France 160,000–area	2 x 3.5 metric tons/hr (2 x 3.9 short tons/hr)	30,000 metric tons (33,000 short tons)	Drying of sewage sludge	FFr 27.50/metric ton ($5.83/short ton)	FFr 32/metric ton ($6.77/short ton)	FFr 59.50/metric ton ($12.60/short ton)
Geneva, Switzerland 340,000–canton	2 x 8.3 metric tons/hr (2 x 9.15 short tons/hr)	115,000 metric tons (126,000 short tons)	Electricity for sale	SF 22.61/metric ton ($6.87/short ton)	SF 41.64/metric ton ($12.70/short ton)	SF 64.25/metric ton ($19.57/short ton)
Kørsor, Denmark 20,000	1 x 2 metric tons/hr (1 x 2.2 short tons/hr)	6,500 metric tons (7,150 short tons)	District heating	Kr 82/metric ton ($12.30/short ton)	Kr 38/metric ton ($5.70/short ton)	Kr 120/metric ton ($18/short ton)
Munich, West Germany 1,315,000	2 x 25 metric tons/hr 3 x 40 metric tons/hr (2 x 27.5 short tons/hr 3 x 44 short tons/hr)	450,000 metric tons (496,000 short tons)	Electricity for sale, district heating	DM 35/metric ton ($11.75/short ton)	DM 12/metric ton** ($4.25/short ton)	DM 47/metric ton** ($16/short ton)

* Only two furnaces were in operation in 1975.

**Average of direct operating costs at North and South plants.

TABLE 7.4: TOTAL CAPACITY AND PER CAPITA CAPACITY

Country	Population (in millions)	Number of Units Installed or Under Construction	Percentage of Units in Europe	Total Installed Capacity (metric tons/hour)*	Total Capacity Installed or Under Construction (metric tons/hour)	Percentage of Total Capacity Installed in Europe	Relative Capacity per Capita (metric tons/hour/no. of inhabitants
Austria	7.5	2	1	55	55	2	7.3
Belgium	9.8	3	1	19	46	1	1.9
Denmark	5.0	45	18	243	277	7	48.2
Finland	4.7	2	1	26	26	1	5.6
France	52.5	29	12	452	494	14	8.6
Greece	9.0	0	0	0	0	0	0
Ireland	3.1	0	0	0	0	0	0
Italy	55.3	16	7	150	250	5	2.7
Luxembourg	0.3	1	–	20	20	1	58.8
Netherlands	13.5	7	3	315	315	10	23.3
Norway	4.0	13	5	12	12	–	3.0
Portugal	8.7	0	0	0	C	0	0
Spain	35.2	3	1	38	46	1	1.1
Sweden	8.2	22	9	183	218	6	22.4
Switzerland	6.5	33	14	255	320	8	39.3
United Kingdom	55.9	9	4	182	192	6	3.2
West Germany	62.0	58	24	1300	1360	40	20.9
Total		**243**	**100**	**3250**	**3631**	**100**	

*To convert metric tons to short tons, multiply by 1.1.

more than one (e.g., Berlin, Munich, Paris, Stockholm, Oslo). There are two exceptions to this pattern, Denmark and Switzerland. In Denmark, several small towns have incineration units. In Switzerland incinerators are centrally located and serve rural areas. The unit at Monthey, Switzerland, for example, generates electricity with energy recovered from the solid waste of 57 villages.

Rated Capacity

Combustion units have been divided into five categories according to capacity:

- less than 5 metric tons/hour
- between 5 and 10 metric tons/hour
- between 10 and 15 metric tons/hour
- between 15 and 25 metric tons/hour
- larger than 25 metric tons/hour

If the average generation of waste in Western Europe is 1 kilogram (2.2 lb) per person, per day, a combustion unit with a capacity of 5 metric tons (5.5 short tons) per hour can serve 100,000 persons in a 20-hour operating day. One would therefore expect to find combustion units with capacities under 5 metric tons per hour in cities under 100,000 inhabitants and those with capacities over 25 metric tons per hour in cities over 500,000 inhabitants. Although only 15% of the units in Western Europe have a capacity over 25 metric tons per hour, they account for 47% of total capacity.

Units under 15 metric tons per hour account for 26% of the total European capacity and 64% of the units. These figures show that, although the very large units handle the bulk of the waste from which energy is recovered, systems for smaller towns (150,000 inhabitants) exist in large number. Energy recovery is not limited to large cities. In fact 34% of all European units have a capacity of under 5 metric tons per hour. Tables 7.5 and 7.6 present information on the sizes of individual furnaces and of combustion plants by country.

In Denmark and Norway, small incinerators predominate. Some burn as little as a quarter of a metric ton per hour. In Germany, France, and the Netherlands, large units are most common.

The largest European plant, located in Rotterdam, the Netherlands, has 6 units of 20 metric tons per hour capacity each. The largest single furnace is located in Paris, France (Ivry II) and has a capacity of 50 metric tons per hour.

In the census of waste-to-energy systems, capacity in terms of metric tons of waste per hour are listed. This measure, rather than yearly output, has been used because different units operate at different percentages of theoretical capacity and for different numbers of hours in a year (depending mostly upon the number of shifts worked). Where daily operating patterns are known, the daily capacity as well has been included.

TABLE 7.5: NUMBER OF INDIVIDUAL FURNACES, BY CAPACITY AND COUNTRY

	Furnace Size (metric tons/hour)*						
Country	**0-5**	**5-10**	**10-15**	**15-20**	**20-25**	**Over 25**	**Total**
Austria		3		2			**5**
Belgium	1	5					**6**
Denmark	35	10	7	1			**53**
Finland		4					**4**
France	13	26	10	4		2	**55**
Italy	18	10	9				**37**
Luxembourg			2				**2**
Netherlands		5	8	4	6		**23**
Norway	12	1					**13**
Spain		1	3				**4**
Sweden	20	8	3	4			**35**
Switzerland	31	9	10	3			**53**
United Kingdom	2	5	12				**19**
West Germany	9	20	37	20	13	5	**104**
Total	**141**	**107**	**101**	**38**	**19**	**7**	**413****

* To convert metric tons to short tons, multiply by 1.1.
**Many units have more than one furnace.

TABLE 7.6: TOTAL CAPACITY BY UNIT SIZE AND BY COUNTRY

	Plant Size (metric tons/hour)*				
Country	**0-5**	**5-10**	**10-15**	**15-25**	**Over 25**
Austria				25	30
Belgium	2			17	27
Denmark	105	46	38	16	72
Finland		10		16	
France	29	76	27	119	233
Italy	10	31	36	94	79
Luxembourg				20	
Netherlands		8	12	21	274
Norway	5	7			
Spain		8	13	25	
Sweden	60	26	12	17	103
Switzerland	45	78	51	146	
United Kingdom	3	13	10	60	106
West Germany	35	78	47	411	789
Total	**294**	**381**	**246**	**987**	**1713**

*To convert metric tons to short tons, multiply by 1.1.

Starting Dates

Energy recovery from waste began, in a small way, in France, Denmark, and Italy before the Second World War. During the 1950s it was introduced in Germany, Sweden, and Switzerland. Other Western European countries first used this technique in the 1960s. The U.K. did not begin until the late 1960s.

The number of units entering service increased from 1960 to 1970 when it peaked. The number entering service has declined slightly since 1970 (see Table 7.7 and Figure 7.11).

FIGURE 7.11: NEW UNITS AND CAPACITY INSTALLED SINCE 1961

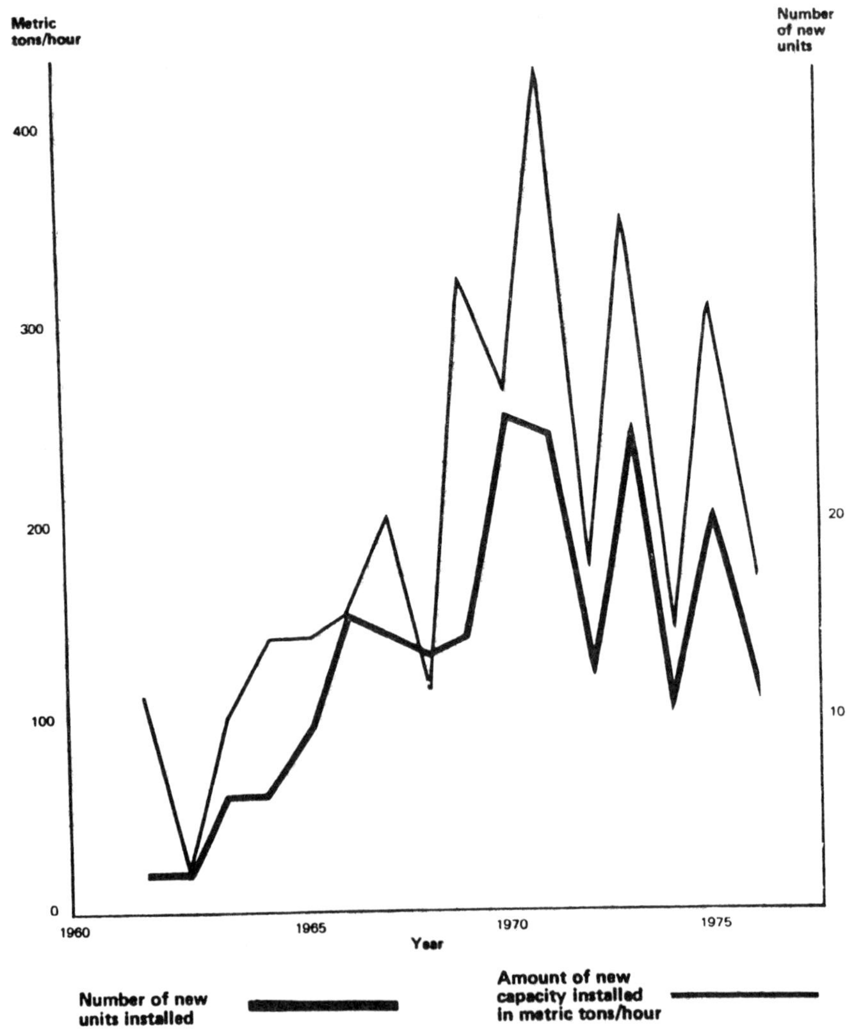

TABLE 7.7: CAPACITY ENTERING SERVICE SINCE 1961
(metric tons/hour)*

	Before 1961	1961	1962	1963	1964	1965	1966	1967	1968	1969	1970	1971	1972	1973	1974	1975	1976	Projected
Austria				25								30						
Belgium	17													2				27
Denmark	24			3		5	13	17	4	5	90	15	18	13	4	28	4	34
Finland			16					10										
France				38	27	68			10	100	44	9	14	18	92	17	5	42
Ireland																		
Italy	20	12						4		11	8		25	68		3		100
Luxembourg																	20	
Netherlands					52				38	64		120	21		12	8		
Norway						**	8		1		1	**	**	**				
Spain																38		8
Sweden	10		4	5		3	13	12	5		18	39	38	25	11			35
Switzerland	33				4	8	16		4	45	12	29	6	17		33	50	65
United Kingdom										8		70	5	27		52	20	10
West Germany	16	100		27	55	54	105	155	48	75	137	111	40	192	19	100	65	60
Total	**120**	**112**	**20**	**98**	**138**	**138**	**155**	**198**	**110**	**308**	**310**	**423**	**167**	**362**	**138**	**279**	**164**	**381**

* To convert metric tons to short tons, multiply by 1.1.
**Fractional amount.

Sources of Information

Most of the information presented on the maps and census tables which follow has been obtained from the grate manufacturers. Where it was impossible to obtain a particular type of information about a system, no entry appears in the list.

MAPS OF WASTE-TO-ENERGY SYSTEMS IN WESTERN EUROPE

FIGURE 7.12: EUROPE

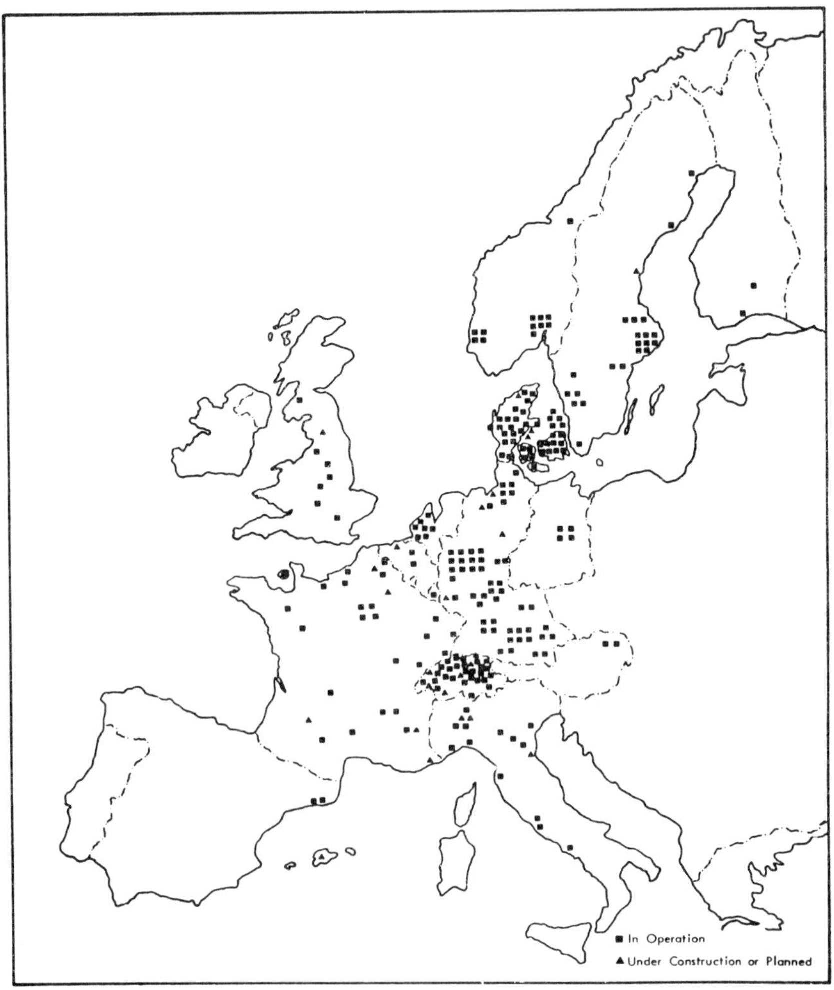

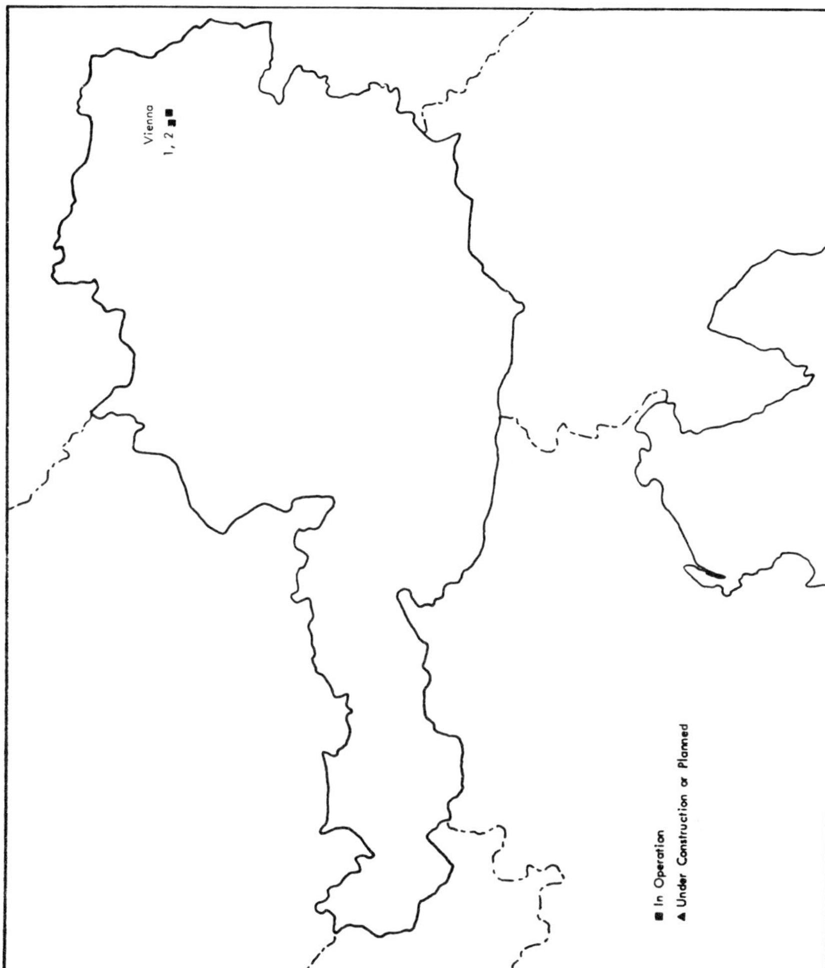

FIGURE 7.13: AUSTRIA

FIGURE 7.14: DENMARK

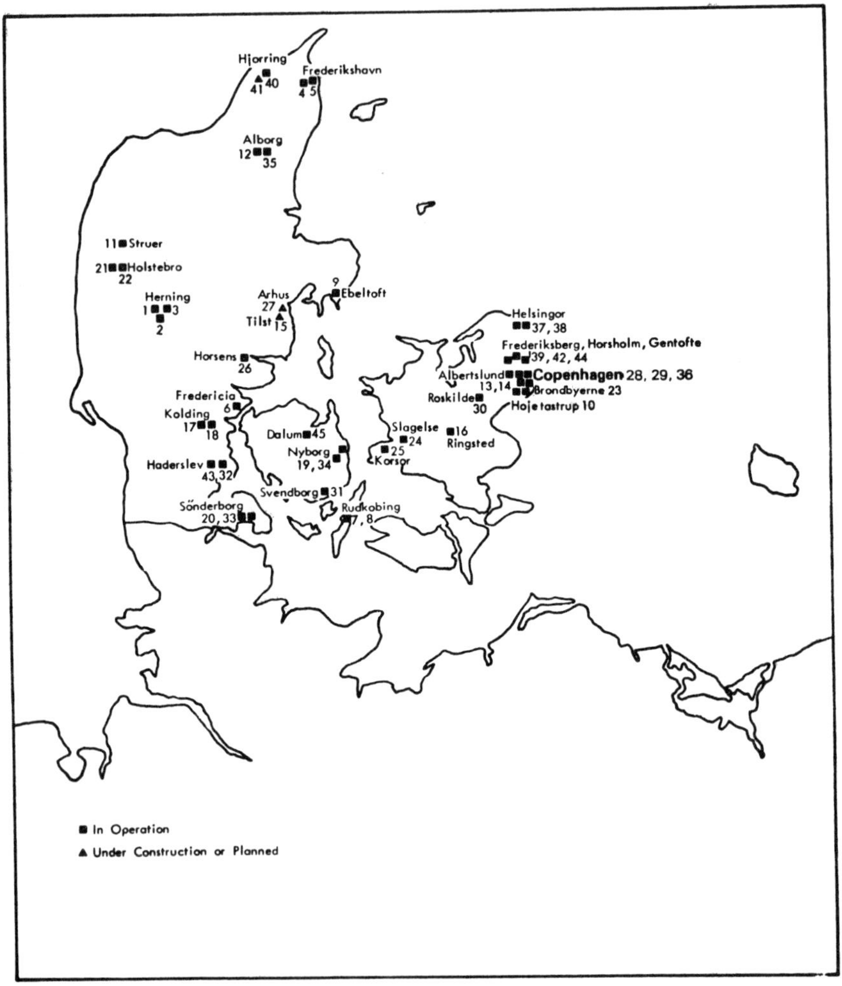

FIGURE 7.15: FRANCE

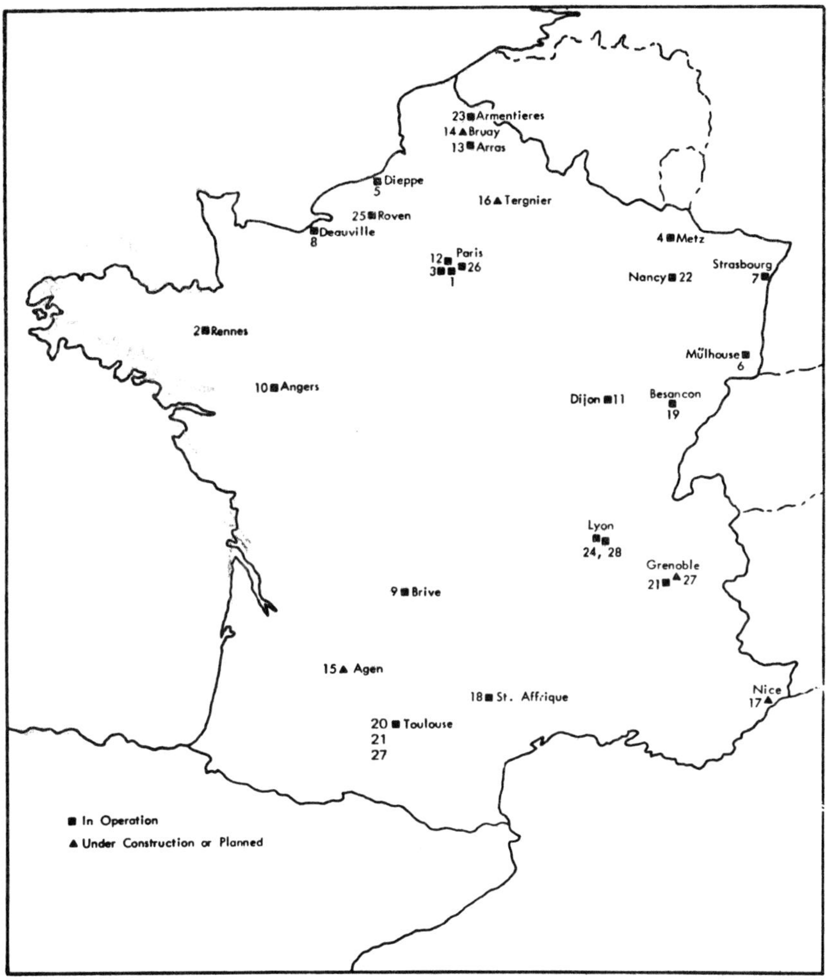

FIGURE 7.16: ITALY

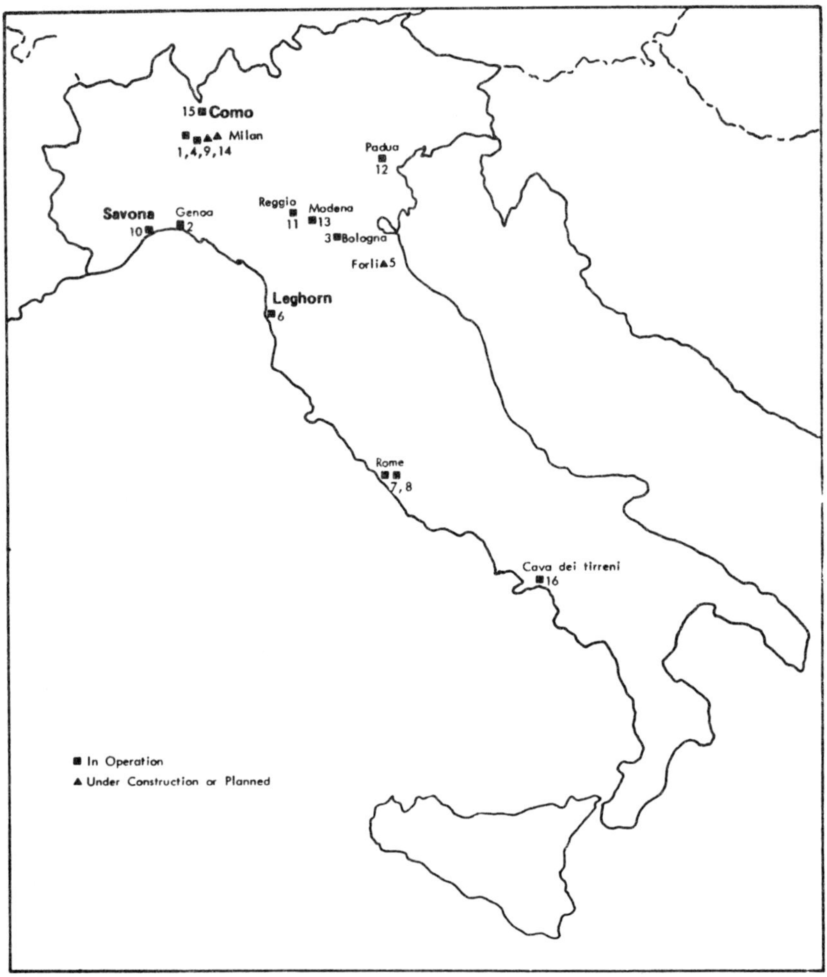

FIGURE 7.17: LUXEMBOURG AND BELGIUM

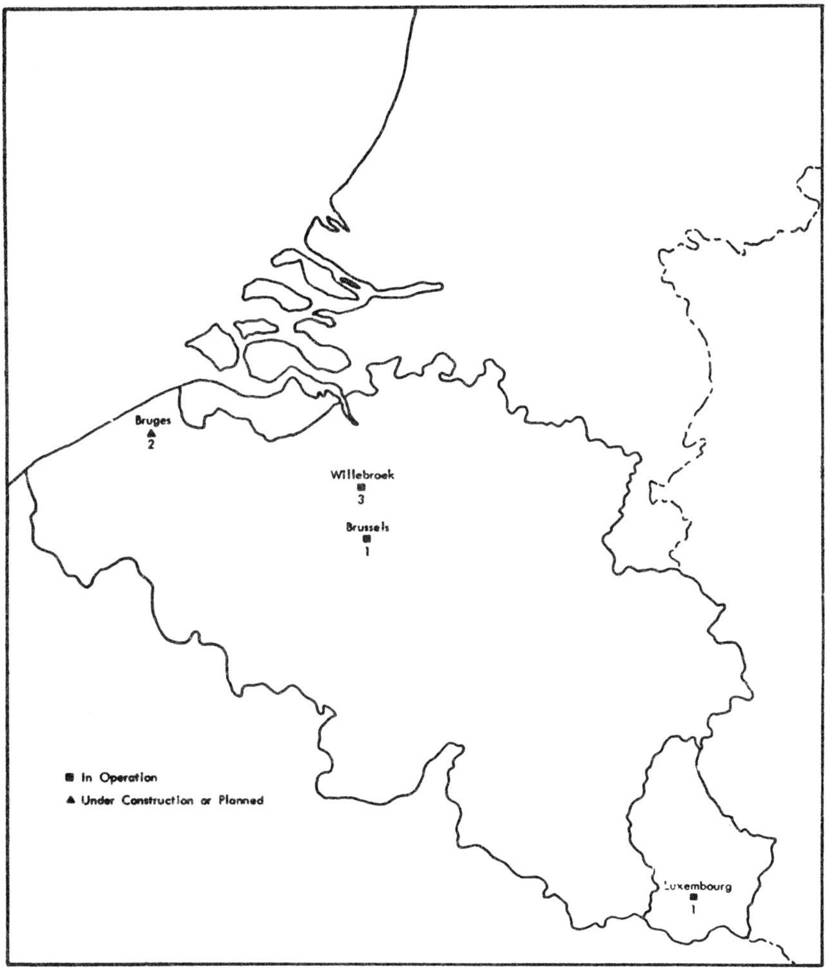

FIGURE 7.18: NETHERLANDS

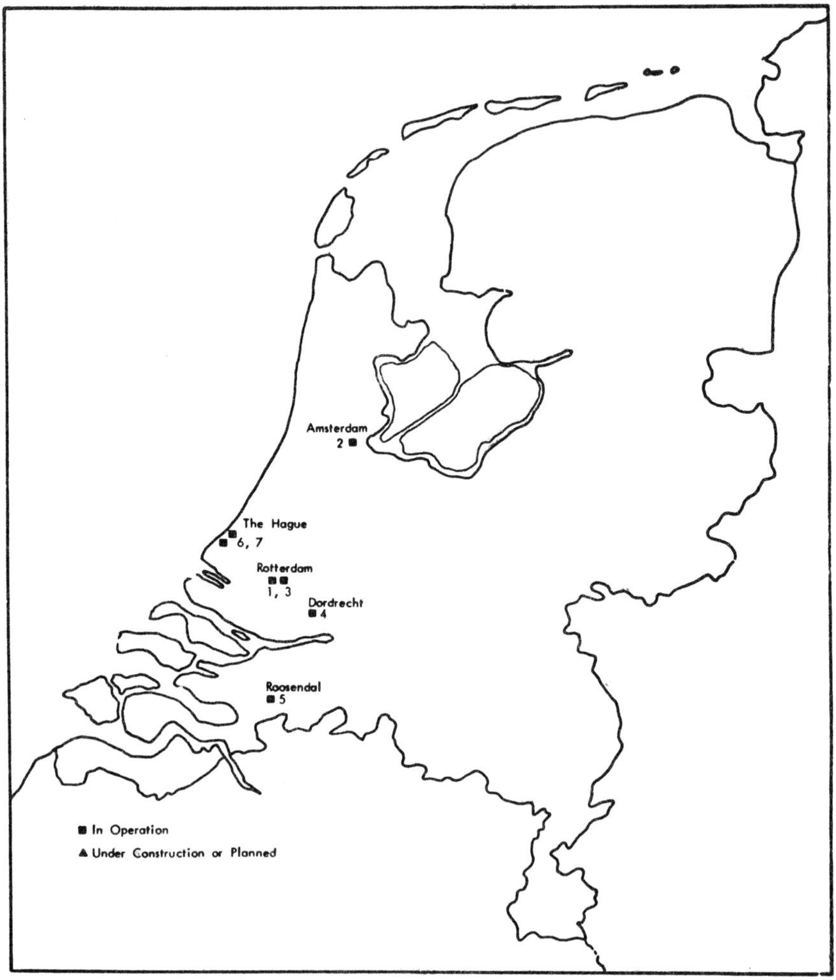

FIGURE 7.19: NORWAY, SWEDEN AND FINLAND

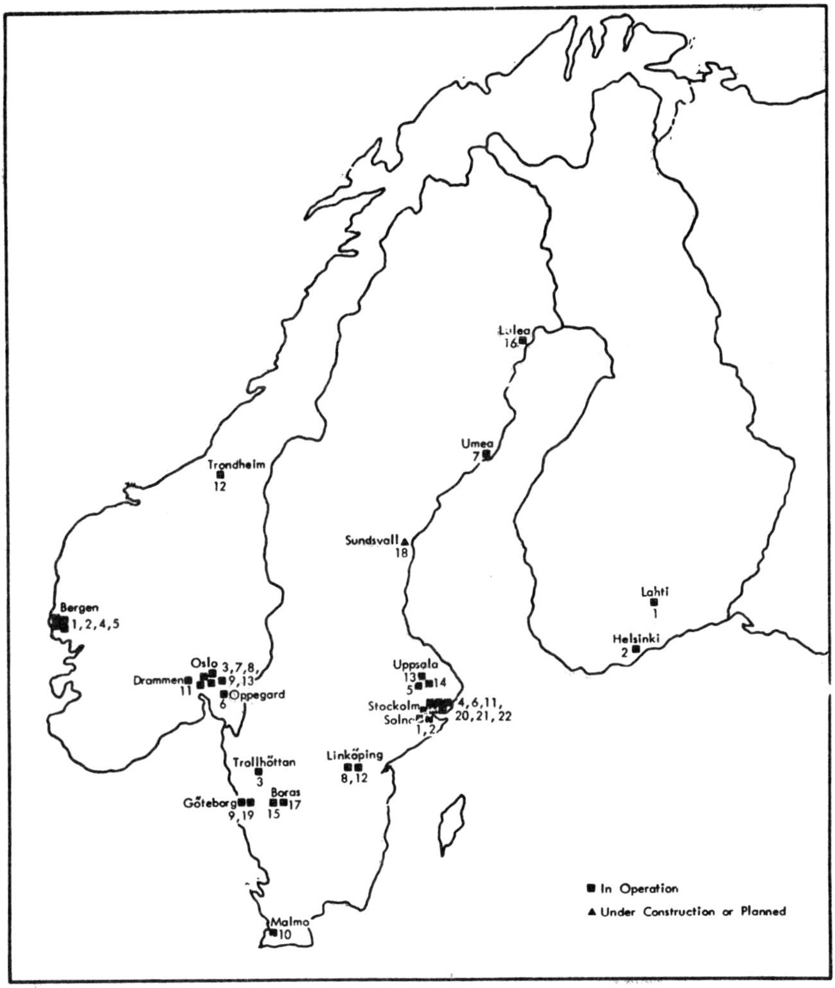

FIGURE 7.20: SPAIN AND PORTUGAL

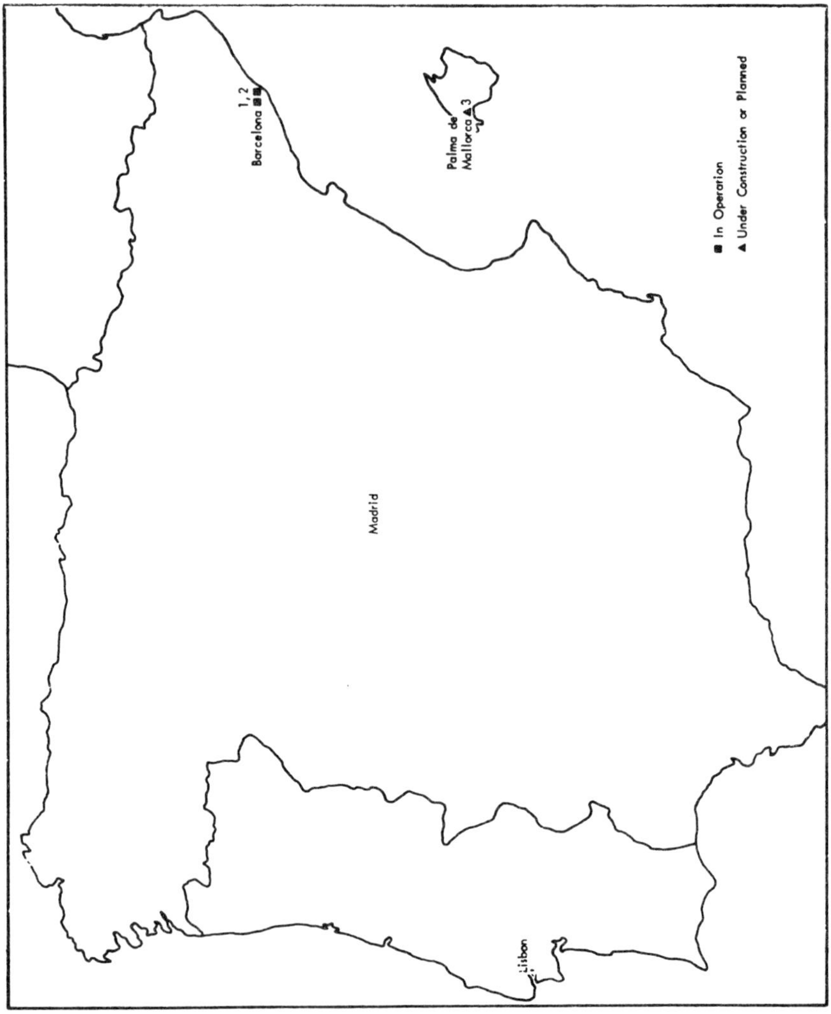

FIGURE 7.21: SWITZERLAND

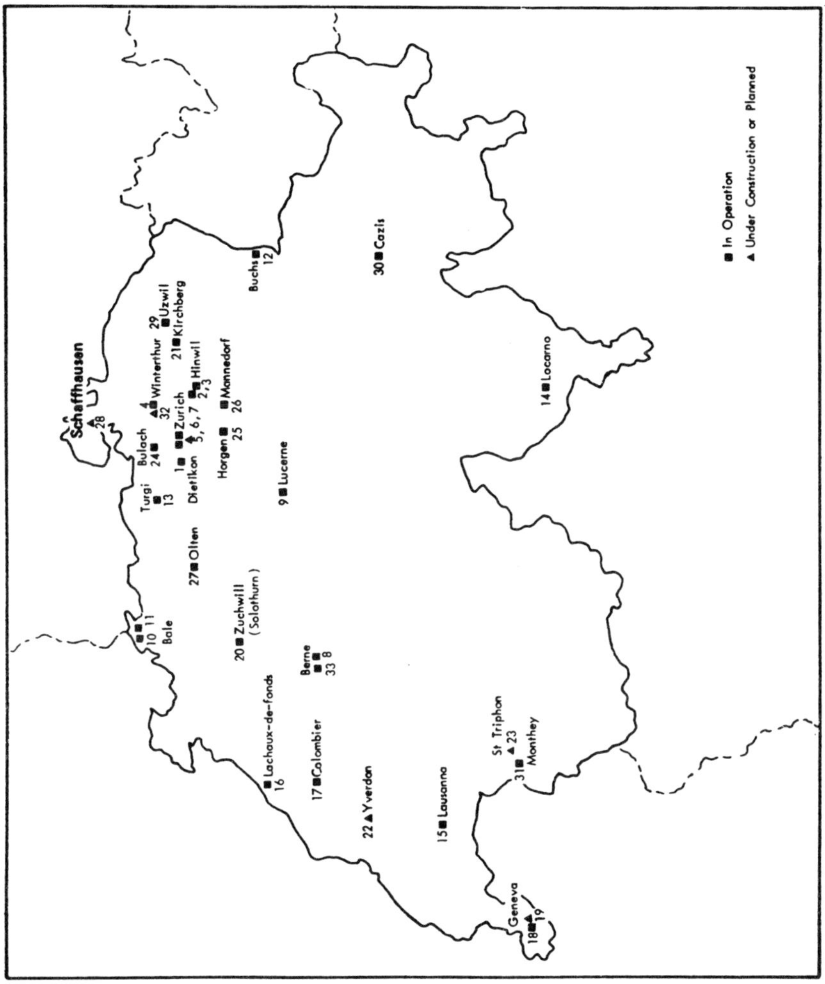

FIGURE 7.22: UNITED KINGDOM

FIGURE 7.23: WEST GERMANY

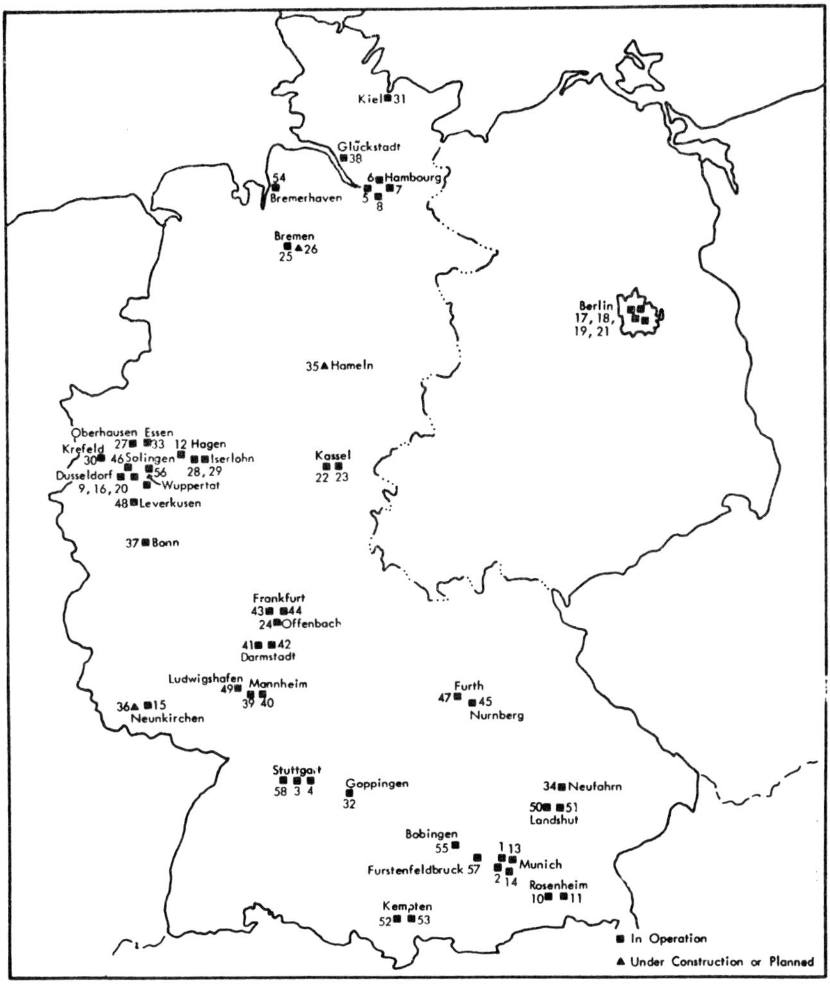

Census of Combustion Units with Energy Recovery in Western Europe

Country	Location/Population	System	Rated Capacity	Energy Recovery	Starting Date	Comments
Austria	1. Vienna 1,600,000	Von Roll	3 x 8.3 t/hr (3 x 200 t/day)	Electricity, hot water, steam for sale and for internal use; hospital heating	1963	
	2. Vienna-Spittelau	Martin	2 x 15 t/hr (720 t/day)	District heating and electricity; steam: 38.6 t/hr per unit	1971	Domestic waste, lower heating value: 1000-1800 k cal/kg
Belgium	1. Brussels 2,000,000	Von Roll	2 x 8.3 t/hr (2 x 200 t/day)	Steam, electricity for sale; steam for internal use	1957	
	2. Bruges 120,000	Martin	3 x 9 t/hr (650 t/day)	Steam: 25 t/hr per unit	under construction	Commercial waste
	3. Willebroek	VKW	1 x 2.3 t/hr	Air preheating	1972 (commissioned)	Industrial waste, lower heating value: 2000 k cal/kg
Denmark	1. Herning 53,000	Bruun & Sørensen	1 x 3 t/hr (1 x 72 t/day)		1963	Domestic waste
	2. Herning extension 1	Bruun & Sørensen	1 x 3 t/hr (1 x 72 t/day)		1971	Domestic waste
	3. Herning extension 2	Bruun & Sørensen	1 x 4 t/hr (1 x 96 t/day)		1973	Domestic waste
	4. Frederkshavn	Bruun & Sørensen	1 x 3 t/hr (1 x 72 t/day)	Hot water, district heating	1965	Domestic waste
	5. Frederkshavn	Bruun & Sørensen	1 x 4 t/hr (1 x 96 t/day)	Hot water, district heating	1971	Domestic waste
	6. Fredericia	Bruun & Sørensen	1 x 4.6 t/hr	Hot water, district heating	1966	Domestic waste
	7. Rudkøbing	Bruun & Sørensen	1 x 1 t/hr	Hot water, district heating	1966	Lower heating value: 2000 k cal/kg
	8. Rudkøbing extension	Bruun & Sørensen	1 x 1 t/hr	Hot water, district heating	1971	Domestic waste

(continued)

Country	Location/Population	System	Rated Capacity	Energy Recovery	Starting Date	Comments
Denmark	9. Ebeltoft	Bruun & Sørensen	1 x 1 t/hr	Hot water, district heating	1967	Domestic waste
	10. Højetastrup	Bruun & Sørensen	1 x 3 t/hr	Hot water, district heating	1967	Domestic waste
	11. Struer	Bruun & Sørensen	1 x 2 t/hr	Hot water, district heating	1967	Domestic waste
	12. Alborg 155,000	Bruun & Sørensen	1 x 5.5 t/hr	Hot water, district heating	1968	Domestic waste
	13. Albertslund	Bruun & Sørensen	1 x 3.5 t/hr	Hot water, district heating	1968	Domestic waste
	14. Albertslund extension	Bruun & Sørensen	1 x 4 t/hr	Hot water, district heating	1974	Domestic waste
	15. Tilst	Bruun & Sørensen	1 x 0.75 t/hr	Hot water, district heating	1968	Domestic waste
	16. Ringsted	Bruun & Sørensen (Flynn and Emrich)	1 x 2 t/hr	Hot water, district heating	1969	Domestic waste
	17. Kolding 53,000	Bruun & Sørensen	1 x 3 t/hr	Hot water, district heating	1969	Domestic waste
	18. Kolding extension	Bruun & Sørensen	1 x 3 t/hr	Hot water, district heating	1972	Domestic waste
	19. Nyborg	Bruun & Sørensen	1 x 3 t/hr	Hot water, district heating	1970	Domestic waste
	20 Sønderborg	Bruun & Sørensen	1 x 3 t/hr	Hot water, district heating	1970	Domestic waste
	21. Holsterbro	Bruun & Sørensen	1 x 3 t/hr	Hot water, district heating	1970	Domestic waste

(continued)

Country	Location/Population	System	Rated Capacity	Energy Recovery	Starting Date	Comments
Denmark	22. Holsterbro	Bruun & Sørensen	1 x 4 t/hr	Hot water, district heating	1975	Domestic waste
	23. Bronsbryerne	Bruun & Sørensen	1 x 4 t/hr	Hot water, district heating	1971	Domestic waste
	24. Slagelse	Bruun & Sørensen	1 x 3 t/hr	Hot water, district heating	1971	Domestic waste
	25. Kørsor 20,000	Bruun & Sørensen	1 x 2 t/hr	Hot water, district heating	1972	Domestic waste
	26. Horsens 53,000	Bruun & Sørensen	1 x 6 t/hr	Sludge drying	1973	Domestic waste
	27. Arhus-Nord 240,000	Bruun & Sørensen	16 t/hr	Hot water, district heating	1977	Domestic waste
	28. Copenhagen-Amager 600,000	Volund	3 x 12 t/hr	Hot water, district heating	1970	Lower heating value: 2500 k cal/kg
	29. Copenhagen West	Volund	3 x 12 t/hr	Hot water, district heating	1970	Lower heating value: 2500 k cal/kg
	30. Roskilde	Volund	2 x 3 t/hr	Hot water, district heating	1966	Lower heating value: 2500 k cal/kg
	31. Svendborg		1 x 3.5 t/hr	Hot water, district heating	1970	Lower heating value: 2000 k cal/kg
	32. Haderslev	Volund	1 x 7.5 t/hr	Hot water, district heating	1967	Lower heating value: 2500 k cal/kg
	33. Sonderborg	Volund	1 x 4 t/hr	Hot water, district heating	1973	Lower heating value: 2500 k cal/kg
	34. Nyborg	Von Roll	1 x 7.5 t/hr	Hot water, steam for internal use	1975	
	35. Aalborg 155,000	Volund	1 x 7.5 t/hr	Hot water, district heating	1972	Lower heating value: 2500 k cal/kg

(continued)

Country	Location/Population	System	Rated Capacity	Energy Recovery	Starting Date	Comments
Denmark	36. Copenhagen West No. 4	Volund	1 x 14 t/hr	Hot water, district heating	under construction	
	37. Helsingør 53,000	Flynn and Emrich	1 x 4.5 t/hr	Hot water, district heating	1967	
	38. Helsingør	Flynn and Emrich	1 x 6 t/hr	Hot water, district heating	1975	
	39. Frederiksberg 100,000	Volund	2 x 6 t/hr	Steam, district heating	1934 (commissioned)	
	40. Hjørring	Volund	1 x 2.5 t/hr	Hot water, district heating	1965 (commissioned)	
	41. Hjørring	Volund	1 x 4 t/hr	Hot water, district heating	1976 (commissioned)	
	42. Hørsholm	Volund	2 x 3 t/hr	Hot water, district heating	1969 (commissioned)	
	43. Haderslev	Volund	1 x 4.5 t/hr	Hot water	1975 (commissioned)	
	44. Gentofte 76,000	Volund	2 x 6 t/hr	Steam	1932 (commissioned)	
	45. Dalum	Volund	1 x 2 t/hr	Hot water	1964 (commissioned)	
Finland	1. Lahti 90,000	Lokomo	2 x 5 t/hr (130 t/day)	District heating: 54,000 MW/h; steam: 11.8 t/hr per unit	1967	

(continued)

Country	Location/Population	System	Rated Capacity	Energy Recovery	Starting Date	Comments
Finland	2. Helsinki 765,000	Von Roll	2 x 8 t/hr (400 t/day)	Hot water for internal use, district heating; electricity: 17.2 GW/h, steam: 15.8 t/hr per unit	1962	
France	1. Paris Issy les Moulineaux 9,000,000	Martin	4 x 15 t/hr (1630 t/day)	Electricity, district heating; steam: 40 t/hr per unit	1965	Industrial waste, lower heating value: 900-2500 k cal/kg
	2. Rennes 180,000	Martin	2 x 5 t/hr (240 t/day)	District heating; steam: 16.2 t/hr per unit	1968	Domestic waste, lower heating value: 1100-2500 k cal/kg
	3. Paris Ivry II	Martin	2 x 50 t/hr (2400 t/day)	Electricity, district heating; steam: 134 t/hr per unit	1969	Industrial waste, lower heating value: 1250-2500 k cal/kg
	4. Metz 165,000	Martin	2 x 6 t/hr (288 t/day)	Electricity, district heating; steam: 18 t/hr per unit	1970	Industrial waste, lower heating value: 1000-2500 k cal/kg
	5. Dieppe 40,000	Von Roll	2 x 2.5 t/hr (2 x 60 t/day)	Sludge drying, internal use	1971	Domestic waste
	6. Mulhouse 200,000	Von Roll	2 x 5 t/hr (2 x 120 t/day)	Steam for sale, district heating, hot water, steam for internal use	1972	Industrial waste
	7. Strasbourg 355,000	Von Roll	3 x 11 t/hr (3 x 300 t/day)	Electricity for sale, district heating, hot water, electricity for internal use	1975	Industrial waste, lower heating value: 1400-1760 k cal/kg
	8. Deauville	Von Roll	2 x 2.5 t/hr (2 x 60 t/day)	Sludge drying	1976	Industrial waste
	9. Brive 160,000	Von Roll	2 x 3.5 t/hr (2 x 75 t/day)	Sludge drying	1974	Industrial waste
	10. Angers 165,000	Von Roll	2 x 5 t/hr (2 x 120 t/day)	Steam for sale and internal use	1975	

(continued)

Country	Location/Population	System	Rated Capacity	Energy Recovery	Starting Date	Comments
France	11. Dijon 185,000	Von Roll	2 x 12.5 t/hr (2 x 300 t/day)	Electricity, steam for sale and internal use	1975	Domestic waste
	12. Bezons Argenteuil	Von Roll	2 x 8.3 t/hr (2 x 200 t/day)	Steam for internal use	1975	Domestic waste
	13. Arras 72,000	Von Roll	1 x 5 t/hr (1 x 120 t/day)	Steam for internal use	1976	Domestic waste
	14. Bruay en Artois 115,000	Von Roll	2 x 5 t/hr (2 x 120 t/day)	Steam for sale	1977	Domestic waste
	15. Agen	Von Roll	1 x 3.125 t/hr (1 x 75 t/day)		1977	Domestic waste
	16. Tergnier	Alberti	2 x 8 t/hr		1972	Sludge drying
	17. Nice-Est 400,000	Martin	1 x 12.3 t/hr 1 x 9.9 t/hr (595 t/day)	Steam: 26 t/hr	1977	Commercial waste
	18. Saint Afrique	Sloppe	1 x 2 t/hr		1972	
	19. Besancon 113,000	CEC	2 x 2.2 t/hr 1 x 3 t/hr (98 t/day)		1971	
	20. Toulouse-Le Mirail 400,000	VKW	2 x 8 t/hr	Electricity, district heating; steam: 21.7 t/hr per unit	1970	Lower heating value: 1000-2200 k cal/kg
	21. Toulouse-Le Mirail	VKW	1 x 8 t/hr	Electricity, district heating; steam: 21.7 t/hr	1975	Lower heating value: 1000-2200 k cal/kg
	22. Grenoble 335,000	VKW	2 x 6.25 t/hr	District heating; steam: 18 t/hr per unit	1971/72	Lower heating value: 1200-2300 k cal/kg
	23. Nancy 260,000	Luchaire	2 x 6.25 t/hr	Steam for sale	1974	
	24. Armentières 55,000	Volund	1 x 3 t/hr		1964	

(continued)

Country	Location/Population	System	Rated Capacity	Energy Recovery	Starting Date	Comments
France	25. Lyon-7ème 1,100,000	Volund	4 x 10 t/hr	Steam, electricity for sale	1963	
	26. Rouen 370,000	Stein	2 x 8 t/hr	Electricity for sale	1967	
	27. Toulouse-Le Ramier				1928	About to be closed down
	28. Paris St. Ouen	Volund	4 x 12 t/hr	Steam for sale	1954	
	29. Grenoble	VKW	6.25 t/hr	District heating; steam: 18 t/hr	1978 (anticipated)	
	30. Lyon	Volund	1 x 10 t/hr	Steam	1973 (commissioned)	
Italy	1. Milan 3,600,000	Volund	2 x 10 t/hr	Steam	1938	
	2. Genoa 875,000	Von Roll	3 x 8.3 t/hr (3 x 200 t/day)	Electricity for sale; electricity, steam for internal use	1972	
	3. Bologna 550,000	Von Roll	3 x 8.3 t/hr (3 x 200 t/day)	Hot water for internal use	1973	
	4. Milan Olgettina	Von Roll	3 x 12.5 t/hr (3 x 300 t/day)	Electricity for sale; electricity, hot water, steam for internal use	1978	
	5. Forli 105,000	Von Roll	2 x 4.15 t/hr (2 x 100 t/day)	Steam for internal use	1977	
	6. Leghorn 173,000	Von Roll	2 x 3.33 t/hr (2 x 80 t/day)	Steam for internal use	1973	
	7. Rome Sorain Cecchini 3,000,000	Von Roll	6 x 4 t/hr	Steam for sale	1973	

(continued)

Country	Location/Population	System	Rated Capacity	Energy Recovery	Starting Date	Comments
Italy	8. Rome Slia Sarr	Alberti Fonsard	2 x 4 t/hr	Steam for sale	1970	
	9. Milan Bruzzano	Volund	3 x 14 t/hr	Steam	under construction	
	10. Savona 120,000	Bartolomeus	1 x 3 t/hr	Steam	1969	
	11. Reggio Emilia 130,000	Bartolomeus	2 x 4 t/hr	Steam	1969	
	12. Padua 265,000	Alberti Fonsard	1 x 12 t/hr	Steam, electricity	1961	
	13. Modena 170,000	Von Roll	2 x 6 t/hr	Steam, electricity	under construction	
	14. Milan Figino	Bartolomeus	2 x 6 t/hr	Steam, electricity	1973	
	15. Como 155,000	Bartolomeus	1 x 4 t/hr	Steam	1967	
	16. Cava Dei Tirreni	Bartolomeus	2 x 1.6 t/hr	Steam	1975	
Luxembourg	1. Luxembourg 100,000	Martin	2 x 10 t/hr (480 t/day)	Electricity for sale and internal use; steam: 25 t/hr per unit	1976	Domestic waste, lower heating value: 1000-2500 k cal/kg
Netherlands	1. Rotterdam Brielselaan 1,125,000	Martin	4 x 13 t/hr (1500 t/day)	Electricity; steam: 30 t/hr per unit	1967	Industrial waste, lower heating value: 1275-2150 k cal/kg
	2. Amsterdam 1,800,000	Martin	4 x 16 t/hr (2160 t/day)	Electricity; steam: 46.5 t/hr per unit	1969	Industrial waste, lower heating value: 1760-2250 k cal/kg

(continued)

Country	Location/Population	System	Rated Capacity	Energy Recovery	Starting Date	Comments
Netherlands	3. Rotterdam	VKW	6 x 20 t/hr	District heating, electricity; steam: 46 t/hr per unit	1970/71 (commissioned)	Domestic waste, lower heating value: 800-3000 k cal/kg
	4. Dordrecht 175,000	Martin	3 x 7 t/hr (504 t/day)	Sludge drying	1972	Industrial waste, lower heating value: 2000 k cal/kg
	5. Roosendaal 50,000	Bruun & Sørensen	2 x 8 t/hr (192 t/day)	Hot water (not in use yet)	1975	Domestic waste
	6. The Hague I 820,000	Von Roll	3 x 12.5 t/hr (3 x 300 t/day)	Electricity, steam for sale; electricity, hot water, steam for internal use	1968	Waste oil recovery
	7. The Hague II	Von Roll	1 x 12.5 t/hr (1 x 300 t/day)	Electricity for sale; electricity, steam, hot water for internal use	1974	
Norway	1. Bergen 235,000	Bruun & Sørensen	1 x 2 t/8hr 1 x 0.25 t/hr		1965	Domestic waste
	2. Bergen	Bruun & Sørensen	1 x 4 t/hr 1 x 0.5 t/hr		1966	Domestic waste
	3. Oslo 720,000	Bruun & Sørensen	1 x 0.5 t/hr		1966	Domestic waste
	4. Bergen	Bruun & Sørensen	1 x 0.25 t/hr		1968	Domestic waste
	5. Bergen	Bruun & Sørensen	1 x 0.25 t/hr		1968	Domestic waste
	6. Oppegard	Bruun & Sørensen	1 x 0.25 t/hr		1968	Domestic waste
	7. Oslo	Bruun & Sørensen	1 x 0.25 t/hr		1968	Domestic waste

(continued)

Country	Location/Population	System	Rated Capacity	Energy Recovery	Starting Date	Comments
Norway	8. Oslo	Bruun & Sørensen	1 x 0.75 t/hr		1968	Domestic waste
	9. Oslo	Bruun & Sørensen	1 x 0.5 t/hr		1970	Domestic waste
	10. Asane	Bruun & Sørensen	1 x 0.5 t/hr		1971	Domestic waste
	11. Drammen 70,000	Bruun & Sørensen	1 x 0.5 t/hr		1972	Domestic waste
	12. Trondheim 130,000	Bruun & Sørensen	1 x 0.5 t/hr		1973	Domestic waste
	13. Oslo	Esslingen	1 x 7 t/hr (1 x 170 t/day)	Electricity; steam. 19.9 t/hr	1966	
Spain	1. Barcelona I 3,165,000	Von Roll	2 x 12.5 t/hr (2 x 300 t/day)	Electricity for sale; electricity, steam for internal use	1975	
	2. Barcelona II	Von Roll	1 x 12.5 t/hr (1 x 300 t/day)	Electricity for sale; electricity, steam for internal use	1975	
	3. Palma de Mallorca 235,000	Volund	1 x 8.25 t/hr (198 t/day)	Steam	under construction	
Sweden	1. Solna 55,000	Bruun & Sørensen	1 x 4 t/hr (1 x 96 t/day)		1973	Domestic waste
	2. Solna	Bruun & Sørensen	8 t/hr (192 t/day)		1974	Domestic waste
	3. Trollhättan 50,000	Bruun & Sørensen	1 x 3.33 t/hr (1 x 80 t/day)	Hot water, district heating	1974	Domestic waste
	4. Stockholm 1,350,000	VKW	2 x 15 t/hr	Electricity; steam: 38.5 t/hr	1970 (commissioned)	Domestic waste, lower heating value: 1600-3000 k cal/kg

(continued)

Country	Location/Population	System	Rated Capacity	Energy Recovery	Starting Date	Comments
Sweden	5. Uppsala 130,000	Bruun & Sørensen	1 x 5 t/hr (1 x 120 t/day)	Steam, district heating	1971	
	6. Stockholm Bollmora	Von Roll	2 x 4.15 t/hr (2 x 120 t/day)	District heating, hot water for internal use	1968	
	7. Umea 60,000	Von Roll	1 x 5 t/hr (1 x 120 t/day)	District heating, hot water for internal use	1970	
	8. Linköping 105,000	Von Roll	1 x 5 t/hr (1 x 120 t/day)	Electricity, district heating, hot water for internal use	1966	
	9. Göteborg 630,000	Von Roll	3 x 12.5 t/hr (3 x 300 t/day)	Hot water for internal use; steam: 42.7 t/hr	1972	
	10. Malmö 300,000	Martin	2 x 8.5 t/hr (408 t/day)	Sludge drying	1973	Industrial waste, commercial waste, lower heating value 3300 k cal/kg
	11. Stockholm Sundbyberg	Volund	2 x 2.5 t/hr (120 t/day)	District heating	1954 (commissioned)	
	12. Linköping	Volund	2 x 2.5 t/hr (1205 t/day)	Steam, electricity, district heating	1958 (commissioned)	
	13. Uppsala	Volund	2 x 2.5 t/hr (120 t/day)	Steam, district heating	1962 (commissioned)	
	14. Uppsala	Volund	1 x 3 t/hr (72 t/day)	Steam, district heating	1965 (commissioned)	
	15. Boras 95,000	Volund	2 x 2.5 t/hr (120 t/day)	Electricity, district heating	1965 (commissioned)	
	16. Lulea 60,000	Volund	2 x 5 t/hr (240 t/day)	Hot water, sludge drying	1969 (commissioned)	

(continued)

Country	Location/Population	System	Rated Capacity	Energy Recovery	Starting Date	Comments
Sweden	17. Boras	Volund	1 x 4 t/hr (96 t/day)	Electricity, district heating	1972 (commissioned)	
	18. Sundsvall Bark 100,000	Volund	2 x 17.5 t/hr	Steam	under construction	
	19. Göteborg	Plibrico	1 x 4 t/hr (96 t/day)	Hot water, district heating	1962	
	20. Stockholm Huddinge	Destructor	1 x 4 t/hr (96 t/day)	Hot water, hospital heating	1971	
	21. Stockholm Solna	Destructor	3 x 4 t/hr (288 t/day)	Hot water, district heating	1967	
	22. Stockholm Bredäng	Destructor	1 x 3 t/hr (72 t/day)	Hot water, district heating	1965	
Switzerland	1. Dietikon 75,000	Sulzer/ Martin	2 x 2.5 t/hr (2 x 45 t/day)	Electricity	1971	
	2. Hinwil Kezo 155,000	Sulzer/ Martin	1 x 5 t/hr (1 x 120 t/day)	Electricity, steam: 12.5 t/hr	1971	Industrial waste, lower heating value: 1100-2200 k cal/kg
	3. Hinwil Kezo	Sulzer/ Martin	2 x 6.25 t/hr (300 t/day)	Electricity, steam: 22.6 t/hr per unit	1976	Commercial waste, lower heating value: 1200-3000 k cal/kg
	4. Winterthur 155,000	Von Roll	2 x 4.15 t/hr (2 x 100 t/day)	Electricity (district heating, electricity, hot water, steam for sale and for internal use)	1965	
	5. Zurich Hagenholz 421,000	Von Roll	2 x 10.8 t/hr (2 x 260 t/day)	District heating (electricity, hot water, steam for sale and for internal use); steam: 28 t/hr per unit	1969	
	6. Zurich Hagenholz II	Martin	1 x 12.5 t/hr (1 x 300 t/day)	District heating; steam: 38.2 t/hr	1973	Commercial waste, lower heating value: 1600-3300 k cal/kg

(continued)

Country	Location/Population	System	Rated Capacity	Energy Recovery	Starting Date	Comments
Switzerland	7. Zurich Josefstrasse	Martin	2 x 7.5 t/hr (1 x 300 t/day)	District heating	1979 (antici-pated)	
	8. Bern 152,000	Von Roll	2 x 4.15 t/hr (2 x 100 t/day)	District heating and industry (electricity, hot water, steam for sale and for internal use)	1954	
	9. Lucerne 173,000)	Von Roll	2 x 4.15 t/hr (2 x 100 t/day)	Electricity (1971), district heating (1976)	1971	
	10. Basel I 460,000	Volund	2 x 8.3 t/hr (2 x 200 t/day)	Industrial heating	1942	
	11. Basel II	Von Roll	2 x 12.5 t/hr (2 x 300 t/day)	Industrial heating	1969	
	12. Buchs 80,000	Alberti Fonsard	1 x 5 t/hr (1 x 120 t/day)	Electricity, district heating, industrial heating	1975	
	13. Turgi 136,000	Alberti Fonsard	2 x 4.15 t/hr (2 x 100 t/day)	Electricity	1970	
	14. Locarno 82,000	Bartolomeus	2 x 2.05 t/hr (2 x 50 t/day)	Industrial heating	1969	
	15. Lausanne 189,000	Von Roll	2 x 4.15 t/hr (2 x 100 t/day)	Electricity, steam, hot water for internal use	1958	
	16. La Chaux de Fond 86,000	CEC	2 x 3 t/hr (2 x 72 t/day)	Electricity, district heating	1972	
	17. Colombier 85,000	Alberti Fonsard	2 x 4.15 t/hr (2 x 100 t/day)	Electricity	1971	
	18. Geneva Cheneviers I 360,000	Von Roll	2 x 8.3 t/hr (2 x 200 t/day)	Electricity for sale; electricity, steam, hot water for internal use	1966	
	19. Geneva Cheneviers II	Martin/Krauss Maffei	1 x 19.15 t/hr (1 x 460 t/day)	Steam: 54 t/hr, electricity	1978 (antici-pated)	

(continued)

Country	Location/Population	System	Rated Capacity	Energy Recovery	Starting Date	Comments
Switzerland	20. Zuchwill 285,000	Von Roll	2 x 10 t/hr (2 x 240 t/day)	Electricity for sale; electricity, steam for internal use	1976	
	21. Kirchberg Bazenheid 71,000	Martin	2 x 3.5 t/hr (2 x 84 t/day)	District heating; steam: 14.7 t/hr per unit	1976	Commercial waste, lower heating value: 1200-3500 k cal/kg
	22. Yverdon 55,000		6 t/hr		under construction	
	23. St. Triphon 165,000		15 t/hr		under construction	
	24. Bülach 65,000	Lurgi	2 x 2.25 t/hr (2 x 54 t/day)	Sludge drying	1969	
	25. Horgen 85,000	Ofag	1 x 3.33 t/hr (1 x 80 t/day)	Internal use	1968	
	26. Mannedorf 55,000	Von Roll	1 x 1 t/hr (1 x 24 t/day)	Internal use	1968	
	27. Olten 73,000	Ofag	1 x 4.15 t/hr (1 x 100 t/day)	Sludge drying	1964	
	28. Schaffhausen 70,000	Von Roll	2 x 2.5 t/hr (2 x 60 t/day)	Internal use	1973	
	29. Uzwil 60,000	Nichols	1 x 3 t/hr (1 x 72 t/day)	Internal use, composting, industrial sludge treatment	1970	
	30. Graubunden 28,000	Ofag	1 x 1.7 t/hr (1 x 41 t/day)	Internal use	1971	
	31. Monthey	Bartolomeus	2 x 3.75 t/hr (2 x 90 t/day)	Electricity	1975	
	32. Winterthur	Martin	1 x 16.7 t/hr (1 x 400 t/day)	Steam: 45 t/hr	under construction	Commercial waste, lower heating value: 1600-3000 k cal/kg

(continued)

Country	Location/Population	System	Rated Capacity	Energy Recovery	Starting Date	Comments
Switzerland	33. Bern II	Von Roll	2 x 10 t/hr (2 x 240 t/day)	Electricity, hot water, steam for sale and for internal use	1975	
United Kingdom	1. Coventry 336,000	Martin	3 x 12 t/hr	Electricity, hot water for industry; steam: 31.5 t/hr per unit	1975	Industrial waste, lower heating value: 1900-2500 k cal/kg
	2. Nottingham 300,000	Martin	2 x 11.7 t/hr (562 t/day)	Electricity, district heating; steam: 24 t/hr per unit	1973	Industrial waste, lower heating value: 1600-2000 k cal/kg
	3. Middleton 55,000	VKW	1 x 8 t/hr	Hot water for internal use	1978 (antici-pated)	Domestic waste, lower heating value: 1000-2000 k cal/kg
	4. Edmonton	VKW	5 x 14 t/hr	Electricity; steam: 39.5 t/hr per unit	1970/71 (commis-sioned)	Domestic waste, lower heating value: 1800-2320 k cal/kg
	5. Linwood	VKW	2 x 8 t/hr	Heat for internal use	1974 (commis-sioned)	Domestic waste, lower heating value: 2720 k cal/kg
	6. Sheffield 500,000	VKW	2 x 10 t/hr	Electricity, district heating	1975 (commis-sioned)	Domestic waste, lower heating value: 2625 k cal/kg
	7. Mansfield 60,000	Bruun & Sørensen	1 x 5.5 t/hr (1 x 132 t/day)	District heating	1972	Domestic waste
	8. Jersey	Martin	1 x 6 t/hr 1 x 4.5 t/hr (294 t/day)	Steam: 17 t/hr per unit	under construc-tion	
	9. Otley/Ilkey	Bruun & Sørensen	3.25 t/hr (1 x 78 t/day)	Hot water, district heating (not yet connected)	1973	Domestic waste
West Germany	1. Munich Nord I 1,400,000	Martin	2 x 25 t/hr (1200 t/day)	Electricity, district heating; steam: 41 t/hr per unit	1964	Commercial waste, lower heating value: 900-1800 k cal/kg (combined waste and coal firing)

(continued)

Country	Location/Population	System	Rated Capacity	Energy Recovery	Starting Date	Comments
West Germany	2. Munich Nord II	Martin	1 x 40 t/hr (960 t/day)	Electricity, district heating; steam: 81 t/hr per unit	1966	Commercial waste, lower heating value: 900-2200 k cal/kg (combined waste and coal firing)
	3. Stuttgart 1,650,000	Martin	1 x 20 t/hr (480 t/day)	Electricity, district heating; steam: 50 t/hr	1965	Commercial waste, lower heating value: 800-2200 k cal/kg
	4. Stuttgart extension	VKW	1 x 20 t/hr	Electricity, district heating; steam: 125 t/hr with oil; 55 t/hr with waste	1970 (commissioned)	Domestic waste, separate furnace with oil firing, lower heating value: 1400-2500 k cal/kg
	5. Hamburg I 1,800,000	Von Roll	2 x 8.3 t/hr (2 x 200 t/day)	Electricity for sale; electricity, heat, steam for internal use	1958	Domestic waste
	6. Hamburg II	Von Roll	3 x 8.3 t/hr (3 x 200 t/day)	Electricity for sale; electricity, heat, steam for internal use	1963	Industrial waste
	7. Hamburg III	Martin	1 x 12 t/hr (288 t/day)	Electricity, district heating; steam: 27 t/hr	1967	Industrial waste, lower heating value: 1200-3190 k cal/kg
	8. Hamburg IV Stellinger Moor	Martin	2 x 19.5 t/hr (1180 t/day)	Electricity; steam: 32-40 t/hr per unit	1973	Domestic waste, lower heating value: 1500 k cal/kg
	9. Dusseldorf 650,000	VKW	4 x 10 t/hr	Electricity, district heating; steam: 16 t/hr per unit	1972	Industrial waste, lower heating value: 800-2000 k cal/kg
	10. Rosenheim	VKW	1 x 4.5 t/hr	District heating, electricity; steam: 7.5 t/hr	1964	Industrial waste, lower heating value: 800-1600 k cal/kg
	11. Rosenheim extension	VKW	1 x 6 t/hr	District heating, electricity; steam: 12 t/hr	1970	Industrial waste, lower heating value: 2000 k cal/kg

(continued)

Country	Location/Population	System	Rated Capacity	Energy Recovery	Starting Date	Comments
West Germany	12. Hagen 200,000	VKW	3 x 6 t/hr	District heating; steam: 16 t/hr per unit	1967	Commercial waste, lower heating value: 800-2000 k cal/kg
	13. Munich South V	Martin	1 x 40 t/hr (960 t/day)	District heating, electricity; steam: 81 t/hr	1969	Commercial waste, lower heating value: 900-2000 k cal/kg
	14. Munich South IV	Martin	1 x 40 t/hr (960 t/day)	District heating, electricity; steam: 81 t/hr	1971	Commercial waste, lower heating value: 900-2000 k cal/kg
	15. Neunkirchen 42,000	Martin	1 x 5 t/hr (120 t/day)	District heating, steam: 175 t/hr	1970	Commercial waste, industrial waste, lower heating value: 2000 k cal/kg
	16. Dusseldorf	VKW	1 x 10 t/hr	Electricity, district heating, steam: 12 t/hr	1965	Domestic waste, lower heating value: 800-2000 k cal/kg
	17. Berlin-Ruhleben 2,000,000	VKW	4 x 16 t/hr	Electricity; district heating, steam: 25 t/hr	1967	Domestic waste, lower heating value: 800-2000 k cal/kg
	18. Berlin-Ruhleben extension 1	VKW	2 x 16 t/hr	Electricity, steam: 26-32 t/hr per unit	1969 (commissioned)	Domestic waste, lower heating value: 800-2500 k cal/kg
	19. Berlin-Ruhleben extension 2	VKW	2 x 16 t/hr	Electricity, steam: 26-32 t/hr per unit	1972 (commissioned)	Domestic waste, lower heating value: 800-2500 k cal/kg
	20. Dusseldorf	VKW	1 x 12.5 t/hr	Electricity; steam: 25-30 t/hr per unit	1972 (commissioned)	Domestic waste, lower heating value: 1000-2100 k cal/kg
	21. Berlin Lichtenberg	VKW	2 x 15 t/hr	Electricity; steam: 32 t/hr per unit	1974 (commissioned)	Domestic waste, lower heating value: 800-1800 k cal/kg

(continued)

Country	Location/Population	System	Rated Capacity	Energy Recovery	Starting Date	Comments
West Germany	22. Kassel 215,000	VKW	1 x 10 t/hr	Electricity; steam: 27.5 t/hr	1968 (commissioned)	Commercial waste, lower heating value: 800-2500 k cal/kg
	23. Kassel extension	VKW	1 x 10 t/hr	Electricity; steam: 27.5 t/hr	1969 (commissioned)	Industrial waste, lower heating value: 800-2500 k cal/kg
	24. Offenbach 120,000	VKW	3 x 10 t/hr	Electricity, district heating; steam: 26 t/hr per unit	1970	Commercial waste, lower heating value: 800-2200 k cal/kg
	25. Bremen 600,000	VKW	3 x 15 t/hr	District heating; steam: 40 t/hr per unit	1970	Industrial waste, lower heating value 800-2000 k cal/kg
	26. Bremen extension	VKW	1 x 20 t/hr	District heating; steam: 88 t/day	1977 (commissioned)	Lower heating value: 3000 k cal/kg
	27. Oberhausen 250,000	VKW	3 x 22 t/hr	Electricity, district heating; steam: 50 t/hr per unit	1972 (commissioned)	Commercial waste, lower heating value: 800-2500 k cal/kg
	28. Iserlohn 60,000	VKW	2 x 8 t/hr	Electricity, district heating; steam: 20 t/hr per unit	1970	Industrial waste, waste oil, lower heating value: 800-2000 k cal/kg
	29. Iserlohn extension	VKW	16 t/hr	Electricity, district heating; steam: 50 t/hr	1973 (commissioned)	Industrial waste, waste oil, lower heating value: 800-2500 k cal/kg
	30. Krefeld 225,000	VKW	2 x 12 t/hr	District heating, sludge drying, electricity for internal use; steam: 38 t/hr per unit	1974 (commissioned)	Industrial waste, waste oil, additional combustion of sewage sludge
	31. Kiel 270,000	VKW	2 x 5 t/hr	District heating; steam: 16 t/hr per unit	1975	Domestic waste, lower heating value: 800-2500 k cal/kg

(continued)

Country	Location/Population	System	Rated Capacity	Energy Recovery	Starting Date	Comments
West Germany	32. Göppingen 50,000	VKW	2 x 12 t/hr	Electricity, district heating; steam: 32 t/hr per unit	1975	Industrial waste, waste oil, lower heating value: 1200-3000 k cal/kg
	33. Essen Karnap 700,000	VKW	5 x 20 t/hr	Electricity; steam: 130 t/hr per unit	1961 (commis-	Industrial waste, lower heating value: 1500 k cal/kg
	34. Neufahrn	VKW	1 x 3 t/hr	District heating; steam: 8 t/hr per unit	1970 (commissioned)	Industrial waste, lower heating value: 1000-2500 k cal/kg
	35. Hameln 50,000	VKW	1 x 10 t/hr	Electricity, steam: 27.2 t/hr	1977	
	36. Neunkirchen 50,000	Martin	1 x 10 t/hr	Steam: 27 t/hr	under construction	Commercial waste, lower heating value: 1500-2500 k cal/kg
	37. Bonn-Bad Godesberg 300,000	Volund	2 x 10 t/hr	District heating, industry	1966	Industrial waste
	38. Gluskstadt	Claudius Peters	1 x 2.5 t/hr	Sludge drying	1963	Domestic waste
	39. Mannheim 330,000	KSG	2 x 12 t/hr	District heating, electricity	1965	Industrial waste, waste oil
	40. Mannheim extension	KSG	1 x 20 t/hr	District heating, electricity	1973	Industrial waste, waste oil
	41. Darmstadt 150,000	Von Roll	2 x 10 t/hr	Electricity, district heating	1967	Industrial waste, waste oil
	42. Darmstadt extension	Von Roll	1 x 12 t/hr	Electricity, district heating	1973	Industrial waste, waste oil
	43. Frankfurt-am-Main I 660,000	Von Roll	2 x 12.5 t/hr	Electricity, steam for internal use and for sale	1966	Commercial waste

(continued)

Country	Location/Population	System	Rated Capacity	Energy Recovery	Starting Date	Comments
West Germany	44. Frankfurt-am-Main II	Von Roll	2 x 12.5 t/hr	Electricity, steam for internal use and for sale	1967	
	45. Nürnberg 500,000	Von Roll	3 x 12.5 t/hr	Steam for sale; electricity, hot water, steam for internal use	1968	Industrial waste
	46. Solingen 180,000	Von Roll	2 x 12.5 t/hr	Electricity, hot water for sale and for internal use	1969	Commercial waste
	47. Fürth 95,000	Von Roll	1 x 3.3 t/hr	District heating; hot water for sale and internal use	1971	
	48. Leverkusen 110,000	Von Roll	2 x 8.3 t/hr	District heating; electricity, steam for internal use	1970	Industrial waste
	49. Ludwigshafen 175,000	Von Roll	2 x 8.3 t/hr	Electricity, district heating, hot water, steam for internal use	1967	Commercial waste
	50. Landshut I 55,000	Von Roll	1 x 3 t/hr	Electricity for sale; electricity, steam for internal use	1971	Industrial waste
	51. Landshut II	Von Roll	1 x 3 t/hr	Electricity for sale and internal use	1974	Industrial waste
	52. Kempten I	Von Roll	1 x 4 t/hr	Steam for internal use	1975	
	53. Kempten II	Von Roll	1 x 5 t/hr	Steam for internal use	1976	
	54. Bremerhaven 150,000	Von Roll	3 x 10 t/hr	Steam for sale; steam, electricity for internal use	1977	
	55. Böbingen	Brule	2 x 1 t/hr	Sludge drying	1975	Commercial waste
	56. Wuppertal 420,000	VKW	4 x 15 t/hr	Electricity for sale and internal use	1976	Industrial waste
	57. Fustenfeldbrück	Keller Peukert	1 x 6 t/hr	Sludge drying	1975	Industrial waste
	58. Stuttgart	VKW	1 x 20 t/hr	Electricity, district heating; steam: 125 t/hr (55 t/hr with waste only)	1965 (commissioned)	Lower heating value: 800-2200 k cal/kg

OTHER TECHNIQUES OF RESOURCE RECOVERY

Besides direct combustion, there are several other processes which recover energy from municipal solid waste. These processes include aerobic fermentation (composting), anaerobic fermentation (generation of methane), the formation of refuse-derived-fuel (RDF), and pyrolysis. In most cases, these techniques are less advanced in Western Europe than in the United States.

Composting has been used in agriculture for ages. As a technique of treating municipal solid waste, composting can yield not only soil conditioner (which should be considered a form of materials recovery) but also energy. The Brikollare Process, developed by Industrie-Werke Karlsrühe Augsburg Aktiengesellschaft (IWKA), is a technique for the joint composting of municipal solid waste and sewage sludge.

In the Brikollare process, the mixture of municipal solid waste and sewage sludge is pressed into briquettes and then palletized into units weighing up to 20 kg (45 lb). The separation of the mixture into these relatively small units promotes fast digestion (about three weeks) and, according to IWKA, eliminates unpleasant odors and harmful bacteria. The dried briquettes can be used as soil conditioner, in pig food, or as a stable element in landfills. IWKA contends that the briquettes can also serve as fuel for combustion units. At present, there is one Brikollare plant in operation in Schweinfurt, West Germany and two more plants are said to be under development.

Anaerobic fermentation producing methane has been used both with municipal solid waste and with animal wastes on farms.

Refuse-derived-fuel is also a substitute for fossil fuel. The processes used to produce RDF do not normally involve fermentation. A process for pelletizing municipal solid waste developed by Buhlers of Zurich, Switzerland is about to be tested in Eastbourne, U.K. This unit, which will have the capacity to produce 7,000 metric tons (7,700 short tons) of RDF per year, may be expanded to produce 40,000 metric tons (44,000 short tons) per year if the first stage is successful.

Public analysis tests on the RDF produced by the Buhlers process indicate that it has about half the calorific value of medium grade coal. Carbonizing the RDF pellets can increase the calorific value to about two-thirds that of high grade coal. RDF is a low sulfur fuel.

The British government's Warren Spring Laboratory has been investigating the production of RDF briquettes but the combustion trials have not been wholly satisfactory. There have also been problems with storage.

Pyrolysis has been investigated by Von Roll in Switzerland and by the Warren Spring Laboratory in England, but no plants have been built. In France, SODETEG conducted some pyrolysis experiments, but the entire project has been abandoned. Pilot plants have been built in Denmark (Destrugas process) and in Sweden (Pyrogas process). A commercial unit using the Andco-Torrax process (10 metric tons per hour) has been built in Luxembourg, but is not yet operating. Another unit is supposed to be built in Grasse, France. A unit is also being built in West Germany which will use the Arcalon process and have a capacity of 24 tons per day.

In addition to energy, a variety of materials can be recovered from municipal solid waste, either through source-separation programs or in front-end separation plants. The recovery of materials from the waste stream is a significant method of conserving both energy and raw materials.

There is considerable interest in Western Europe in front-end separation techniques. Research is being carried out in France (BRGM), Sweden (Flakt and Motala companies), West Germany (Krauss-Maffei), and England (Warren Spring Laboratory). The Flakt company has been operating a separation plant (capacity 30,000 metric tons per year) in Sweden since the beginning of 1975 and has contracted to build a full scale commercial plant (capacity 120,000 metric tons per year) in the Netherlands. Krauss-Maffei operates a pilot plant in Munich, West Germany and Cecchini operates one in Rome, Italy.

Appendix

CONVERSION TABLES

English to Metric Conversions

To Convert	Multiply By	To Obtain
ton (short)	0.9078	ton (metric)
feet	0.3048	meters
square feet	0.0929	square meters
cubic feet	0.0283	cubic meters
cubic yards	0.7646	cubic meters
gallons	0.003785	cubic meters
square miles	2.59	square kilometers
atmospheres	1.0133	bars
British thermal units	0.252	kilocalories
British thermal units	2.52×10^{-7}	kilotherms
British thermal units	1055	joules

Metric to English Conversions

To Convert	Multiply By	To Obtain
ton (metric)	1.102	ton (short)
meter	3.281	feet
square meter	10.76	square feet
cubic meter	35.31	cubic feet
cubic meter	1.308	cubic yards
cubic meter	264.2	gallons
square kilometer	0.386	square miles
bar	0.987	atmospheres
kilocalorie	3.9685	British thermal units
kilotherm	3.97×10^{6}	British thermal units
joule	9.486×10^{-4}	British thermal units
kilowatt-hour	1.1	Therms

CURRENCY CONVERSION TABLES

The source of all of the following currency conversion tables was "International Financial Statistics," International Monetary Fund.

French Francs to U.S. Dollars

Year	F/$
1969	5.20
1970	5.53
1971	5.51
1972	5.04
1973	4.45
1974	4.81
1975	4.29
1976	4.78

Swiss Francs to U.S. Dollars

Year	SF/$
1963	4.32
1964	4.32
1965	4.32
1966	4.33
1967	4.33
1968	4.30
1969	4.31
1970	4.31
1971	4.11
1972	3.82
1973	3.17
1974	2.98
1975	2.58
1976	2.50

Danish Kroner to U.S. Dollars

Year	Kr/$
1969	7.52
1970	7.50
1971	7.41
1972	6.98
1973	6.28
1974	5.65
1975	6.18
1976	6.05

West German Marks to U.S. Dollars

Year	DM/$
1960	4.17
1961	4.00
1962	4.00
1963	3.98
1964	3.98
1965	4.00
1966	3.98
1967	4.00
1968	4.00
1969	3.69
1970	3.64
1971	3.48
1972	3.19
1973	2.67
1974	2.59
1975	2.46
1976	2.52

Sources Utilized

The following reports were used in the preparation of this book:

CONS/2103-1

European Waste-to-Energy Systems, Case Study of the Thermal Complex of Toulouse-Le Mirail (France), Resource Planning Associates, Inc., Washington, D.C. and Paris, May 1977.

CONS/2103-2

European Waste-to-Energy Systems, Case Study of Geneva-Cheneviers (Switzerland), Resource Planning Associates, Inc., Washington, D.C. and Paris, and Consultex, SA, Geneva, May 1977.

CONS/2103-3

European Waste-to-Energy Systems, Case Study of Korsor, Denmark, Resource Planning Associates, Inc., Washington, D.C. and Paris, May 1977.

CONS/2103-4

European Waste-to-Energy Systems, Case Study of Munich: Munich North IA & IB, Munich North II, and Munich South IV & V, Resource Planning Associates, Inc., Washington, D.C. and Consultex, SA, Geneva, May 1977.

CONS/2103-5

European Waste-to-Energy Systems, Case Study of the Thermal Complex of Brive, France, Resource Planning Associates, Inc., Washington, D.C. and Paris, June 1977.

CONS/2103-6

European Waste-to-Energy Systems, an Overview, Resource Planning Associates, Inc., Washington, D.C. and Paris, June 1977.

PB-270 219

European Developments in the Recovery of Energy and Materials from Municipal Solid Waste, W. David Conn, University of California, Los Angeles, California 90024, May 1977.

FLUIDIZED BED COMBUSTION OF COAL AND WASTE MATERIALS 1977

by Lee Yaverbaum

Pollution Technology Review No. 35
Energy Technology Review No. 15

Coal represents about 85% of the U.S. fossil fuel resources and will certainly play an important role in achieving greater energy independence.

Fluidized bed combustion (FBC) has a number of attractions for getting energy from coal—it is a more compact, cheaper method of burning high-sulfur coal and other fuels in an environmentally acceptable manner.

In the basic concept, coal, which is crushed but not ground, is fed with dolomite or limestone into a fluidized bed. Water, pumped through coils in the bed, is converted into steam for electricity, process heat or other application. Temperature in such a fluidized bed reactor is maintained at several hundred degrees lower than on conventional boiler grids. Operation at elevated pressures offers further advantages—the hot gas from a pressurized system can be expanded through a gas turbine, thereby increasing the power generating efficiency even further.

This technology can be used to substitute many low grade fuels directly for coal. This book gives details for burning coal tailings, municipal refuse and sludge, industrial and agricultural waste materials, oil shale and petroleum fractions. With further development of the technology, energy contributions from fluidized bed combustion will be very substantial indeed. Most of the data in this book are based on federally funded studies—a partial and condensed list of contents follows here.

ISBN 0-8155-0671-6 **267 pages**

ENERGY FROM BIOCONVERSION OF WASTE MATERIALS 1977

by Dorothy J. De Renzo

Energy Technology Review No. 11
Pollution Technology Review No. 33

One of the chief gaseous products of the anaerobic decomposition of organic matter is methane, CH_4. This is how natural gas was formed in prehistoric times along with other fossil fuels.

By applying this principle today in environmentally acceptable fashion it is possible to bioconvert municipal solid sewage, animal manure, agricultural and other organic wastes into substitute natural gas (95% CH_4). In its simplest essentials the process consists of loading the material into a digester (a closed tank with a gas outlet). Given favorable thermal and chemical conditions, the appropriate biological processes will then take their course.

The bioconversion of waste materials to methane provides at least partial solutions not only to the energy problem, but also to the solid waste disposal problem. The harvesting of heretofore undesirable vegetations, such as algae, water hyacinths, and kelp as "energy crops" offers unconventional opportunities for supplementary utilization of natural resources.

This book describes practical methods for the bioconversion of waste matter. It is based on reports of academic and industrial research teams working under government contracts. A partial and condensed table of contents follows here. Chapter headings and important subtitles are given.

Note: Each chapter is followed by bibliographic reference lists in order to provide the reader with easy access to further information on these timely topics.

ISBN 0-8155-0656-2 **223 pages**

WASTE TREATMENT WITH POLYELECTROLYTES AND OTHER FLOCCULANTS 1977

by Sidney J. Gutcho

Pollution Technology Review No. 31

The recycling of wastewater to a purity acceptable and useful in population centers and industry can be accomplished through the judicious use of polyelectrolytes and other flocculants, such as natural latexes.

Polyelectrolytes are high molecular weight polymers and are classified into three types—anionic, cationic and nonionic. They promote the process whereby suspended solids and colloidal materials in the water are agglomerated into masses sufficiently large to settle. Being polyelectrolytes, they neutralize any static charges on dirt particles, allowing them to come together.

Correctly applied, organic flocculants are practically nontoxic, biodegradable, small in volume, easily incinerated, and effective under varied pH and temperature conditions. This makes them ideal for clarification of water from reservoirs, lagoons, etc. They are being used at an increasing rate for the removal of solids from various industrial wastewaters including the effluents of the mining and papermaking industries.

This book presents a broad spectrum of the manufacturing technologies and practical applications, as they have been elaborated by many industries and municipalities. A partial and condensed table of contents follows here. Numbers in parentheses indicate the number of processes per topic.

ISBN 0-8155-0648-1 **274 pages**

FUEL CELLS FOR PUBLIC UTILITY AND INDUSTRIAL POWER 1977

Edited by Robert Noyes

Energy Technology Review No. 18

Fuel cells are generators of electricity containing no moving parts except small extraneous pumps for the movement of fuel and oxidant into the cell and the products of oxidation out of the cell.

Public utilities and industrial consumers require high voltage, three-phase alternating current. In this application fuel cells must compete with turbine-driven generators which provide such current. While the output of a fuel cell is low voltage DC power, cells may be connected in various series and parallel arrangements to give whatever voltage is desired, but mechanical rotary converters or delicate electronic inverters must then provide conversion to AC (60 cycles for each phase in the USA).

The advantages of a fuel cell system over turbine-driven generators lie in greater efficiency at full load which even increases as the load diminishes, so that inefficient peaking generators are not needed. There are considerable pollution control advantages to be gained as well. Because the various suitable fuels react electrochemically rather than by burning in air, no nitrogen oxides are formed. For the same reason, emissions of unburned or partly burned gaseous and particulate products are practically nil.

Fuel cell installations using inverters have a long life with relatively little maintenance. It is now entirely feasible to have small, completely unattended fuel cell power plants using waste fuels on location (such as hydrogen from chlorine production) to produce convenient power.

This book, based on information derived from U.S. government-contracted studies, contains considerable practical down-to-earth technical information relating to fuel cells for power plants. A partial and condensed table of contents follows.

ISBN 0-8155-0676-7 **325 pages**

RADIOACTIVE WASTE DISPOSAL 1977

Low and High Level

Edited by W. R. Gilmore

Pollution Technology Review No. 38

The increasing use of nuclear energy for electric power generation, as well as the expanding applications of radioisotopes in various fields, are inevitably associated with growing amounts of radioactive wastes.

This book describes the technology being developed to concentrate and confine both high level and nonhigh level radioactive wastes so that they may be stored or otherwise gotten rid of in a comparatively safe and compact mode according to governmental guidelines.

Present radioactive waste management practices at U.S. nuclear facilities and commercial power plants are outlined in detail. Careful attention is also given to proposed schemes for radioactive waste management resulting from advances in this particular technology.

Most important is a long-term solution to the problem of accumulation of wastes from the nuclear fuel cycle. Processes are under development to convert the wastes into solid and virtually insoluble forms (the so-called vitrification processes). There are also various proposals for permanent storage ("disposal") of solidified wastes under the ocean bed or in deep geological formations where they will be virtually inaccessible.

This book is based mainly on federally-financed studies. The Table of Contents (partially reproduced here in condensed form) is organized in such a way as to serve as a subject index. A valuable glossary and list of abbreviations of nuclear and technical terms serves to familiarize the reader with this timely topic.

ISBN 0-8155-0684-8 **356 pages**

HOW TO SAVE ENERGY AND CUT COSTS IN EXISTING INDUSTRIAL AND COMMERCIAL BUILDINGS 1976

An Energy Conservation Manual

by Fred S. Dubin, Harold L. Mindell and Selwyn Bloome

Energy Technology Review No. 10

This manual offers guidelines for an organized approach toward conserving energy through more efficient utilization and the concomitant reduction of losses and waste.

The current tight supply of fuels and energy is unprecedented in the U.S.A. and other countries, and this situation is expected to continue for many years. Never before has there been as pressing a need for the efficient use of fuels and energy in all forms.

Most of the energy savings will result from planned systematic identification of, and action on, conservation opportunities.

Part I of this manual is directed primarily to owners, occupants, and operators of buildings. It identifies a wide range of opportunities and options to save energy and operating costs through proper operation and maintenance. It also includes minor modifications to the building and mechanical and electrical systems which can be carried out promptly with little, if any, investment costs.

Part II is intended for engineers, architects, and skilled building operators who are responsible for analyzing, devising, and implementing comprehensive energy conservation programs. Such programs involve additional and more complex measures than those in **Part I.** The investment is usually recovered through demonstrably lower operating expenses and much greater energy savings.

A partial and much condensed table of contents follows here:

PART I

PART II

Much of the technology required to achieve energy savings is already available. Current research is providing refinements and evaluating new techniques that can help to curb the waste inherent in yesteryear's designs. The principal need is to get the available technology, described here, into widespread use.

ISBN 0-8155-0638-4 **$24** **725 pages**

COAL ASH UTILIZATION
FLY ASH, BOTTOM ASH AND SLAG
1978

Edited by S. Torrey

Pollution Technology Review No. 48

Ash is a waste product left after the burning of many combustible substances, and fly ash is the accepted term for the finely divided residue that results from the combustion of ground coal. It is easily disseminated by flue gases, unless checked and collected by suitable devices.

Fly ash particles are primarily composed of silica and alumina. Secondary ingredients are carbon and oxides of iron, calcium, magnesium and sulfur.

Energy and environmental considerations over the coming years point to greater use of coal: As more utilities are forced to shift from gas and oil to coal as a source for fuel, the available quantities of fly ash will increase. This book shows how millions of tons of fly ash and other ashes can be disposed of annually in an environmentally acceptable manner. There is, for instance, a well-developed technology using lime + fly ash aggregates in pavements, buildings and bridge construction. Since these aggregates generally require only 2 to 5 percent lime, the expenditure of money and energy for these mixes is very low, thus making an attractive case for the expanding use of lime + fly ash aggregates.

The book is based mostly on federally funded studies as listed in the bibliography at the end of the volume. A partial and condensed table of contents follows here.

ISBN 0-8155-0722-4 **370 pages**

COGENERATION OF STEAM AND ELECTRIC POWER 1978

Edited by Robert Noyes

Energy Technology Review No. 29

The cogeneration concept, as presented in this book, refers to the generation of both process steam and electricity on site by heating water with a primary fuel, such as coal, oil or gas.

When fuel was cheap, industries were not interested in capital outlays necessary for cogeneration equipment. Now, however, the picture has changed. Refineries, chemical plants, paper mills and other big energy users can get maximum value from their fuel by a "topping cycle," viz. producing steam and running it through a turbine generator before using the steam for processing operations. Also, cogeneration can be accomplished through a "bottoming cycle," utilizing process steam prior to electrical power generation. Both methods are discussed in detail.

In the past few years, the U.S. Government has been placing much emphasis on the need for conserving energy. One of the ways of saving considerable amounts of energy is the development of cogeneration plants by those industries which use significant amounts of electricity and thermal energy for process heat and/or space heating. Since electric power generation produces heat as a by-product, any thermal energy which can be used produces a fuel savings of from 10 to 30%, depending on the applications and methods involved. The studies indicate that for the greatest potential development of cogeneration, government support may be required.

Various technological studies, such as material prepared for the National Science Foundation and the Federal Energy Administration, were the basis for this review. A partial, condensed table of contents follows here.

ISBN 0-8155-0706-2

258 pages